Palgrave Studies in Professional and Organizational Discourse

Titles include:

Cecilia E. Ford
WOMEN SPEAKING UP
Getting and Using Turns in Workplace Meetings

Sue Garton and Keith Richards (*editors*)
PROFESSIONAL ENCOUNTERS IN TESOL
Discourses of Teachers in Teaching

Rick Iedema (*editor*)
THE DISCOURSE OF HOSPITAL COMMUNICATION
Tracing Complexities in Contemporary Health Care Organizations

Louise Mullany
GENDERED DISCOURSE IN THE PROFESSIONAL WORKPLACE

Keith Richards
LANGUAGE AND PROFESSIONAL IDENTITY
Aspects of Collaborative Interaction

H. E. Sales
PROFESSIONAL COMMUNICATION IN ENGINEERING

Forthcoming titles:

Jonathan Crichton
THE DISCOURSE OF COMMERCIALIZATION
Exploring Language in Organizational Ecology

Palgrave Studies in Professional and Organizational Discourse
Series Standing Order ISBN 978–0–230–50648–0
(*outside North America only*)

You can receive future titles in this series as they are published by placing a standing order. Please contact your bookseller or, in case of difficulty, write to us at the address below with your name and address, the title of the series and the ISBN quoted above.

Customer Services Department, Macmillan Distribution Ltd, Houndmills, Basingstoke, Hampshire RG21 6XS, England

Language and Professional Identity

Aspects of Collaborative Interaction

Keith Richards
University of Warwick, UK

First published in hardback 2006
This paperback edition published 2009 by
PALGRAVE MACMILLAN

Palgrave Macmillan in the UK is an imprint of Macmillan Publishers Limited,
registered in England, company number 785998, of Houndmills, Basingstoke,
Hampshire RG21 6XS.

Palgrave Macmillan in the US is a division of St Martin's Press LLC,
175 Fifth Avenue, New York, NY 10010.

Palgrave Macmillan is the global academic imprint of the above companies
and has companies and representatives throughout the world.

Palgrave® and Macmillan® are registered trademarks in the United States,
the United Kingdom, Europe and other countries.

ISBN-13: 978–1–4039–3800–8 hardback
ISBN-13: 978–0–230–58011–4 paperback

This book is printed on paper suitable for recycling and made from fully
managed and sustained forest sources. Logging, pulping and manufacturing
processes are expected to conform to the environmental regulations of the
country of origin.

A catalogue record for this book is available from the British Library.

A catalog record for this book is available from the Library of Congress.

10   9   8   7   6   5   4   3   2   1
18   17   16   15   14   13   12   11   10   09

Transferred to Digital Printing in 2009

Dedicated to the memory of
William Hayes and Roger Mann
Teachers
Bablake School, Coventry

With gratitude

Contents

List of Figures and Tables

Figures

Table

Acknowledgements

Over the years I have benefited immeasurably from the insights of friends and colleagues whose influence permeates this book in ways that are too subtle and too pervasive to specify. I am particularly grateful for the abiding friendship of Paul Rogers and all the insights – personal as well as professional – that this has brought over many very precious encounters. Julian Edge, Sue Garton and Steve Mann, colleagues and friends for many years, have shared with me their discoveries and insights, deepening my own understanding and enthusiasm as well as bringing enjoyment to my work. I should also like to thank my editor, Jill Lake, for her understanding and encouragement. My deepest debt, as ever, is to my long-suffering wife, Marie, and to my daughters, Francesca and Louisa, who have made life even more interesting.

Transcription Conventions

Transcription is based on Jefferson 1979, but conventional orthography is used wherever possible and the transcription of laughter is more crudely represented than is standard in conversation analysis (a note on this can be found on page 94). Pauses are timed to tenths of seconds up to one second, then in half-second intervals. Italicised speaker name indicates uncertain attribution.

(0.6)	Pause	So (0.6) what are we going to do?
(.)	micropause	Put it (.) away
[ ]	Overlap	A: He saw it ⌈and⌉ stopped B: ⌊oh⌋
[[	Simultaneous start	A: ⌈⌈And the B: ⌊⌊So she left it behind
=	Contiguous utterances	A: We saw her yesterday= B: =And she looked fine
.	Falling intonation	That was foolish.
,	Continuing contour	I took bread, butter, and jam
?	Rising intonation	Who was that?
!	Animated tone	Look!
↓↑	Marked rise or fall in intonation	We had a ↑ wonderful time
—	Emphasis	Put it a<u>way</u>
italics	Uttered with laughter in voice	And there he was *behind the door*
CAPS	Louder than surrounding talk	There are TWO OF THEM
° °	Quieter than surrounding talk	But it was lost °so he said°
> <	Quicker than surrounding talk	And >quick as that< she was off
-	Sound cut-off	All over the pl- the floor
:	Sound stretching	We waited for a lo:::ng time
(xxx)	Unable to transcribe	We'll just (xxxxxxxxx) today
(word)	Unsure transcription	And then he (juggled) it
(( ))	Other details	Leave it alone ((moves book))

1
Introduction

Introduction

Identity is an issue. Historically, we have moved to a point where even the process of constructing identity is laid bare for public dissection: newspaper columnists analyse the effectiveness with which politicians and their PR teams establish and refine particular identities for public consumption, popular television programmes expose for watching millions the attempts of individuals in goldfish bowl settings to construct identities designed to elicit popular support, and representations of corporate identity are pressed upon us as customers and employees. The process of construction itself has entered a new level of consciousness as the Web, offering multiple dimensions of representation, creates for all of us the possibility of establishing different identities, each valid within its particular domain. Even the process of collecting data which might help us to understand these phenomena is challenged by the impenetrability of the represented self (Atkinson and Silverman, 1997; MacLure, 1993) and the mutability of whatever might lie behind it:

> The central feature of the self in modern society is its reflexivity, a constant questioning and reconstruction of the self in a lifetime project. We are constantly constructing and revising our personal stories and so reconstructing ourselves.
>
> (Craib, 1998: 2)

Far from turning the researcher's knees to jelly, this apparent indeterminacy offers interesting avenues of exploration. If construction and reconstruction are indeed fundamental to the lived self, the process by which this is achieved assumes a particular importance and offers a legitimate focus for investigation (see, for example, Antaki and Widdicombe, 1998; Malone, 1997). But it is also the case that, as Bolinger (1975: 333) has observed, '[t]here is no limit to the ways in which human beings league themselves together for self-identification, security, gain, amusement, worship, or any of the other purposes that are held in common' and that the groups thereby formed have their own distinctive identities, formed and nurtured through the alchemy of human inter- action. The processes involved are as yet little understood and this book offers an insight into the ways in which such groups are interaction- ally constituted. Its particular concern is with the group rather than the individual, and specifically with the collaborative professional group. But in exploring the ways in which group identity is established and maintained in such contexts, the following chapters suggest that the interactional pattern upon which this depends may also sow the seeds of potentially destructive 'grouptalk'.

The importance of mutually supportive teams, in work as elsewhere, is generally acknowledged, and considerable efforts have been invested in developing cultures in which they can flourish. When I first began the research upon which this book is based it was with a view to throwing light on the nature of such teams and the interactional foundations of their success. What I did not predict when I began was the imminent dissolution of all three of the demonstrably successful groups making up my sample. Why was one group made redundant, another broken up and a third placed in a position where all chose to leave the parent insti- tution within less than a year? I knew it would be foolish to look at any single cause, but were there interactional clues that I had at first failed to recognise? My search led me beyond the borders of linguistic study and into the realms of group psychology where the concept of 'group- think' is used to explain a way of behaving that lead otherwise successful groups to make disastrous decisions. Such decisions must be reached through talk and the patterns in this talk will not emerge overnight; they will already be part of the group's distinct interactional identity, developed through innumerable encounters, casual and formal, planned and ad hoc, extended and fleeting. Where might traces of such patterns be detected in the everyday talk of such groups and are they a feature of the evolution of all collaborative enterprises, an ineradicable aspect of group identity that might one day undermine its very constitution? The answers to this must remain speculative, but the paradox at its heart is inescapable: the interactional processes that contribute to the success of the group within a wider context may reinforce the group's own identity at the expense of external affiliations, leaving it isolated and vulnerable.

The remainder of this chapter will outline key themes of the book, beginning with an overview of issues in individual and group identity. This will be followed by an introduction to the study of workplace interaction and the issues associated with it. The section following this, which focuses on the concept of backstage talk, highlights the interactional context of the study and the orientation of its central argument. The chapter concludes with a discussion of research issues, an introduction to the groups at the heart of the project and a sketch of the chapters to follow, the first of which will take up the major theme of group identity and its construction.

Individual and group identity

Identity, whether group or individual, is never merely a matter of assuming or assigning a label; it is something that is formed and shaped through action. As with individuals, the identity of a group might be described in any number of ways, both formal and informal, private and public: the professional designation, 'computer section', might be at the same time 'that bunch of bloody reactionaries down the corridor' and 'the ones we depend on to stay in business'. The labels will change according to circumstance, and over time some of the standard descriptions may be revised as the group redefines itself through its relations with others and as the dynamic process of identity (re)construction continues. The pervasive presence and significance of such construction is reflected in the increasing popularity of Membership Categorisation Analysis (e.g. Sacks, 1992a and b; Hester and Heglin, 1997; Psathas, 1999; Lepper, 2000), which explores how membership of particular categories (e.g. 'doctor', 'mother', 'caller') is made relevant in talk through the use of Membership Categorisation Devices (MCDs) and related rules.

The focus in this study is how collaborative professional groups construct and reaffirm their own identities, and even though the steps by which such identities were originally developed are now lost, traces of them will be detectable in the talk of the group, in the ways its members orient to one another and to those outside. Such traces are evidence of the inescapable fact that every time we speak we reveal – whether deliberately or accidentally – something of ourselves and who we take ourselves to be. In fact, the semiotic resources at our disposal are so rich and subtle that our command of them at the conscious level is necessarily limited.

Sometimes a single word or action is enough to represent a specific identity, with all its social and interactional associations, as Margaret Thatcher revealed during her period as the UK prime minister. Appearing before the assembled press on the steps of a hospital where her daughter-in-law had just given birth to a child, she announced in tones fit for the most solemn of state occasions, 'We are a grandmother!' The use of the plural form of the first person was neither accidental nor, since this is a form reserved for the reigning monarch, insignificant. Coming at a time when Thatcher's behaviour was becoming increasingly regal and was recognised as such by contemporary satirists, this statement confirmed a view of herself that offended many but surprised few. She also provided a wonderful example of how our linguistic choices help to define us, our situation and our relationship to those we are addressing, though of course such personal characterisations are not binding on others – I saw no evidence of people throwing themselves onto one knee and crying "Vivat Regina!" in response to her announcement.

Although in Thatcher's case a single pronoun uttered by an individual represented a bid for a particular identity, more than one person can work together over extended interactional space to construct a shared identity. In Extract 1.1, twin sisters in their 80s who have lived together all their lives are recalling their childhood and youth and have reached the point where their sister, Fanny, took over the family shop. Notice how Mabel's use of 'dad' (not 'my dad') embraces both of them and how they build their turns on one another's, repeating the same words and phrases, overlapping and interrupting, so that the description emerges as a shared account. Although they are separate individuals, they talk as one:

Extract 1.1

```
001  Mabel:   it happened ⌈that
002  Doris:              ⌊it happened=
003  Mabel:   =because all- Fanny went into it first 'cos
004           dad had a stroke
005           (0.5)
006  Doris:   ⌈He had a stroke after⌉ twelve months.
007  Mabel:   ⌊IN the post office.  ⌋
008  Doris:   After twelve months ⌈he had a stroke
009  Mabel:                       ⌊He was fifty-four.
```

I choose twin sisters as a paradigm case of a shared interactional identity and would not expect to find the same degree of congruence in a professional group. However, as Chapter 2 demonstrates, such groups can act as one in order to preserve their ways of doing things. The power of established routines and identities was brought home to me many years ago when I visited my daughter's primary school on the 'new parents afternoon' and found myself sitting alongside other parents in a tiny chair that thrust my knees up to my chin. Just before the head teacher arrived, his deputy provided a short introduction, bending forward and addressing us in her best classroom voice. Having provided all the relevant information about the forthcoming meeting, she ended by adding, 'And if any of you need to *go*, there's a ... (leaving a gap where the key word would have appeared, but compensating by pointing towards the staff lavatory and dropping her voice to an emphatic whisper) over *there*.' I found myself, in the company of all the other parents, nodding slowly and deliberately, with all the meek solemnity of a five-year-old.

There is a danger, though, in making straightforward assumptions about role definition and assigning identity simply on the basis of this. Even in the most clear cut of situations, local contingencies may override default positions. For example, an exchange on the subject of digestive problems between a doctor and a patient in a hospital consulting room manifests the key elements of a professional encounter: setting (the consulting room), topic (digestive problems), and the identity of the participants (doctor and patient). An examination of the talk will reveal features identifying it as an example of doctor–patient interaction and other elements characteristic of client–professional encounters in general. However, those involved in the talk are not actors on a stage whose every word is predetermined. Responsibility for the dialogue that eventually emerges falls to them, and although the process of construction must be based on a shared understanding of the interactional business in hand, which in turn depends on the key elements identified above, these elements are not determinants of the talk so much as resources on which the interactants can draw. It may be that at some points in the encounter neither participant orients to these features, in which case the talk will take a very different turn. Schegloff provides a telling illustration of the potential faultlines:

> So the fact that a conversation takes place in a hospital does not ipso facto make technically relevant a characterization of the setting, for a conversation there, as 'in a hospital' (or 'in the hospital'); it is the talk of the parties that reveals, in the first instance *for them*, whether or when the 'setting in a/the hospital' is relevant (as compared to 'at work,' 'on the east side,' 'out of town,' etc.). Nor does the fact that

the topic of the talk is medical ipso facto render the 'hospital setting' relevant to the talk at any given moment. Much the same points bears on the characterization of the participants: For example, the fact that they are 'in fact' respectively a doctor and a patient does not make those characterizations ipso facto relevant (as is especially clear when the patient is also a doctor); their respective ages, sex, religions, and so on, or altogether idiosyncratic and ephemeral attributes (for example, 'the one who just tipped over the glass of water on the table') may be what is relevant at any point in the talk. On the other hand, pointed used of technical or vernacular idiom (e.g. of 'hematoma' as compared to 'bruise') may display the relevance to the parties of precisely that aspect of their interaction together. It is not, then, that some context independently selected as relevant affects the interaction in some way. Rather, in an interaction's moment-to-moment development, the parties, singly or together, select and display in their conduct which of the indefinitely many aspects of context they are making relevant, or are invoking, for the immediate moment.

(Schegloff, 1987: 219)

Schegloff's account offers a powerful illustration of the dangers inherent in the assumption that 'being doctor' necessarily involves consistently 'actively being doctor', but it does so on the basis of a further simplification. For when the doctor knocks over the water he or she does not become 'the person who knocked over the water' but 'the doctor who knocked over the water'. The difference, from an interactional perspective, may be profoundly significant. How I deal with someone who knocks over the water is never determined by the simple act of knocking over the water. Leaving aside the many potentially significant influences deriving from setting and circumstance (knocking over a plastic glass of water at a picnic is an act hardly comparable with the tumbler that spills water on the vital page at a public signing ceremony), the role of the participants may also be brought into focus through subsequent talk: my own childhood memories of the difference between *my* spilling tea on the tablecloth and a similar action on the part of an important guest are, I would think, not significantly different from those of other people.

Schegloff is right to claim that we are never, by virtue of our officially sanctioned role in a particular setting, simply 'the doctor', but by the same token we are rarely *simply* 'the person who spilt the water'. Such characterisations are descriptively convenient but ontologically incomplete, and adequate analysis depends on the irreducibility of interactional identity. Sensitivity to the ways in which different identities might be constructed, invoked, exploited, or otherwise made relevant develops from a perspective that sees identity as an interactional resource rather than a category in search of a label.

From this perspective the concept of group identity is neither more nor less problematic than individual identity; both have to be worked for by participants and the same analytical procedures can be used to expose the processes by which they are established. It would be disingenuous to claim that, as a label, 'collaborative group X' carries the same descriptive load as 'doctor', but this is merely a matter of general familiarity. They are both resources, and it may even be that the very availability of the latter and the ease with which it can be invoked embed it so deep within the interaction that its roots are less easily dug out by the analyst than the more exposed process by which collaborative identity is worked up and maintained. In any case, between 'the doctor' and 'the person who knocked over the water' – the general category and the local categorisation – lies the interesting territory of the professional group in a particular setting, where we find the creation of local categories ('the one who knocks things over') with general reference within the group (jokes alluding to this characteristic). This is what makes the study of particular groups so fascinating, and potentially so important for our understanding of workplace interaction.

Workplace interaction – backstage perspectives

Exchanges such as that between a doctor and patient, where those involved must find procedurally relevant ways of speaking in order to achieve particular goals, provide a natural site for the investigation of what constitutes 'institutional' talk. It is therefore hardly surprising that client–professional, or, more broadly, insider–outsider exchanges have featured prominently in research on such talk. Interactionally, these encounters present a rich vein for the analyst to mine, but they are only one aspect of a larger institutional reality and interest in them led until recently to a relative neglect of the interactional construction of the professional world itself, prompting calls for a realignment of priorities (e.g. Sarangi and Roberts, 1999). Contact with patients, although important, is only one aspect of a doctor's professional life, for example; behind it lies an institutional identity reflexively constructed through continual interaction with other individuals and groups within relevant working contexts. Fundamental to such contexts is the existence of groups, and the membership of these groups, constructed and represented through talk, is inextricably bound up with professional identity, albeit represented in ways that may be very different from those evidenced in encounters with outsiders.

This shift of attention to the interaction of professional groups raises interesting analytical issues that are reflected in developments in the field. Perhaps the most fundamental question, addressed by Drew and Heritage (1993) in the introduction to their seminal collection on talk at work, is that of the relationship between ordinary conversation and specifically institutional interaction. It is generally recognised that simple distinctions between the two are untenable, but the ways in which institutional realities are invoked or made relevant by the parties involved are traceable in their exchanges. Drew and Heritage draw particular attention to goal orientations, special and particular constraints on what may be allowable, and specific inferential frameworks, while a later discussion (Drew and Sorjonen, 1997) focuses on orientation to institutional identities through person reference, lexical choice, grammatical forms, turn-taking and institutionally specific inferences. As clients or representatives, for example, we design our language in order to achieve particular institutional goals and in so doing we make relevant in the talk different aspects of a broader institutional reality. The particularities of this design will vary from institution to institution and from encounter to encounter, but there is clear evidence that particular linguistic phenomena can be tracked through their use in a range of different institutional settings (see, for example, Drew, 2003 on formulations).

Although features such as these allow analytical purchase, it would be a mistake to think that analysis is simply a matter of identifying institutional ways of speaking. As Arminen (2000) notes, if we want to understand what goes on in institutions we must direct attention to the way in which the constitution of the talk reveals the institutional resources that are used to perform relevant tasks. This calls for a view of institutional interaction that not only captures the particularities of a given situation but also connects with wider organisational norms and practices. Miller captures the relationship well:

> Institutional discourses consist of the fundamental assumptions, concerns, and vocabularies of members of settings and their usual ways of interacting with one another. Institutional discourses are shared and standardized frameworks for anticipating, acting in, and reflecting on social settings and interactions. They allow and

constrain setting members to organize their interactions as instances of standardized types of social relationships and produce conditions for responding to issues in predictable ways.

(Miller, 1994: 282-3)

Traditionally, such connections have been sought in ethnographies, but increasing interest in the way organisational realities are realised in particular instances of talk and behaviour has shifted attention away from the explanation of macro-social categories (Idema and Wodak, 1999: 14). An increasing focus on the reflexive relationship between emergent local activities and workplace discourses (Cicourel 2003) has allowed researchers greater freedom in the nature and extent of their investigations, exposing some of the subtle interplay between social and institutional activity. Koester's (2004) study of the relationship between relational and transactional exchanges, for example, while rather categorical in terms of its basic distinction, offers valuable insights into the range and placement of relational sequences. Other research projects have collected rich and varied data covering a range of everyday workplace interactions (e.g. Stubbe, 1998), bringing the analytical microscope to bear on exchanges that are part of organisational life but not part of formal business (e.g. Holmes, 2000a).

This opening up of perspectives is gradually broadening our understanding of the complex ways in which workplace talk is constitutive of the business of 'being at work' and revealing the wide range of interactional resources upon which participants draw. The chapters that follow explore this landscape, exposing the interactional forces generated by three different professional groups as they go about their daily institutional business. Unsurprisingly, the interactional resources on which they draw are familiar, but the ways in which these are utilised and the strength of their binding powers are in many ways surprising, and though the ending of the story is not a happy one the same cannot be said of the groups themselves.

Before introducing characters and theme, some delineation of setting and scene is called for and in this respect Goffman's distinction between front and back stage activities is fundamental. In his discussion of the presentation of self (1959/1971: 109), Goffman introduces the concept of a *region*, 'any place that is bounded to some degree by barriers of perception' and draws an important distinction between a *front region*, where the professional performance (e.g. a doctor consulting with a patient) takes place, and a *back region*, where considerations of 'decorum' applying to the front region are suspended and 'the suppressed facts make an appearance'. It is, he suggests, 'a place, relative to a given performance, where the impression fostered by the performance is knowingly contradicted as a matter of course' (ibid.: 114). In fact, his claims go deeper than this, suggesting that within western society there exist two fundamentally different behavioural languages associated with the two settings, and he selects a pungent list of back region behaviours in order to make his point:

> The backstage language consists of reciprocal first-naming, cooper-ative decision-making, profanity, open sexual remarks, elaborate griping, smoking, informal dress, 'sloppy' sitting and standing posture, use of dialect or sub-standard speech, mumbling and shouting, playful aggressivity and 'kidding', inconsiderateness for the other in minor but potentially symbolic acts, minor physical self-involvements such as humming, whistling, chewing, nibbling, belching, and flatulence. The frontstage behaviour language can be taken as the absence (and in some sense the opposite) of this.
>
> (Goffman 1959/1971: 129)

Goffman's provocative definition of frontstage behaviour, expressed here in terms of its opposition to this rather baroque collection of human antics, is entirely in keeping with the theatrical metaphor around which he weaves his arguments. However, his dependence on this metaphor has attracted criticism (e.g. Burns, 1992), and the harshly contrastive relationships between front and back stage and between team and audience are not well-suited to representing the complexities of back region interaction in all its forms. An even more fundamental problem arises if Goffman's characterisation is treated as representing the essential nature of what takes place backstage, which is certainly what his use of the term 'consists of' suggests. In organisations where the front office is just one among many departments or where encoun-ters with clients take place in the field, back regions may be extensive, embracing different degrees of separation of insiders from clients and producing a complex range of interactional behaviours the interplay of which is not reducible to crude categories, however colourful. A single meeting, for example, might involve characteristically back region beha-viour playing off serious engagements where business is enacted and where 'audience' members are championed by members of the team. A new characterisation is therefore needed, one that responds to this complexity while recognising the important relationship captured by Goffman.

The first step in such a characterisation must be to distinguish front and back stage from front and back region, treating the former in their everyday sense as physical settings and using regions to represent specific behavioural constructions within these settings, thus preserving the essentially contrastive relationship between front and back regions. This position is similar to the one advanced by Sarangi and Roberts (1999: 20), who recognise that for the purposes of mapping the relevant territory the most useful starting point is Goffman's distinction between front region and back region activities, but then go on to develop their own representation of relevant research activity in terms of front and back stage. They propose distinguishing *frontstage* studies, insider/outsider encounters of the sort described in the doctor–patient encounter, from *backstage* investigations where the focus is on the ways in which the institutional world and professional knowledge are constructed. A crude but not altogether unhelpful distinction between these two might lie in a difference of emphasis between the ways in which the professional world is *constructed* backstage and *represented* frontstage – although the relationship is of course a reflexive one and, as we shall see in Chapter 7, the construction of insider/outside interaction will itself be represented backstage as part of the construction of the professional world.

The central argument of this book does not depend on the relationship between front and back stage as such, but in focusing on ways in which backstage territory is marked out and occupied it demonstrates how participants draw on representations of frontstage realities as a resource in their construction of collaborative identity. The importance of the wider organisational context was not lost on Goffman, who remarks how 'individuals attempt to buffer themselves from the deterministic demands that surround them' (1959/1971: 116). Such buffering is, of course, not always uncontested; Hargreaves, for example, has drawn attention to administrative colonisation in some schools, where teachers' rights to use backstage time as they wish have been eroded by principals who insist that preparation time and personal activity – be it business or recreational – do not mix. He also puts his finger on a key feature of backstage freedoms: they 'foster informal relations that build trust, solidarity and fellow-feeling' (1994: 110), the more formal business of the institution benefiting from the 'interpersonal platform' that results from this. What he does not go on to do is explore how this platform is constructed. Inspection of this reveals a complex interactional picture that may not be entirely positive.

Analysis

In order to understand, let alone explain, why workplace communicative practices are the way they are we, as analysts, need to
immerse ourselves in specific workplace contexts. And shedding
light on specific and local practice is equally important if we want
to go beyond understanding and explaining to contributing to
change.

(Sarangi and Roberts, 1999: 2)

Nothing happens of itself; somewhere there is a history, somehow there
are consequences. Although talk is shaped in the moment, it is not of
the moment; its bedrock is laid down over countless exchanges and
the judgements that inform its microinteractional realisation draw on
profound reserves of experience and understanding. Hence it does not
yield easily to analysis, and the researcher must somehow establish
a working compromise between a desire to draw general conclusions
and the responsibility to do justice to the uniqueness of the particular.
In representational terms, this means choosing between a large collection of short extracts illustrating 'finished' features of interaction and a
much narrower selection of longer texts exposing the processes through
which these features are created. The former offers the benefits of instant
accessibility and range of description but at the expense of convincing
detail, while the latter provides a richer and more compelling representation of the particular at the expense of breadth of vision – although,
as Cicourel (1968) points out, a particular case may reveal something of
all social interaction. The analysis presented in the chapters that follow
leans heavily towards the latter, sacrificing the advantages of accumu-
lated instances in order to lay bare the process of construction-in-action
that produces the distinctive characteristics of collaborative professional
groups. Over the course of a book the gradual accumulation of individual cases to some extent redresses the balance, but this is merely an
incidental benefit: the relevant challenge lies not in providing eviden-
tial bulk but in using an analytical scalpel to peel away interactional
layers that for the most part have been left undisturbed by previous
research.

The natural tool for such work is to be found in the tradition of
conversation analysis (CA):

Conversation analysis of social encounters provides highly detailed
understandings of how participants use available interpretive and

interactional resources to construct, sustain and change concrete social relationships and settings.

(Miller, 1997: 157)

The development of CA as a distinct discipline is associated with a series of lectures delivered by Harvey Sacks in the 1960s, but the influence of the founder of ethnomethodology, Harold Garfinkel, is profound. Garfinkel's overriding interest was in the pervasiveness of commonsense knowledge, and the ways in which this is utilised to maintain a shared understanding of the world around us and the activities in which we are involved. His position, crudely put, is that relevant categories are those developed by the actors in social situations as part of a dynamic process of situated knowledge and shared understandings (Garfinkel, 1967).

Building on this work and Goffman's contributions to our understanding of the interaction order, Sacks demonstrated that ordinary conversation offers a unique insight into the ways in which people understand and represent their social world (Sacks, 1992 a and b). The emphasis in the analysis of conversation is not on how interactants obey the relevant rules, but on how they jointly construct the conversation and their shared understanding of what is happening in it. Attention is therefore directed to the sequential development of the talk, how each turn relates to what has gone before and looks forward to what will follow. Nothing is considered in isolation and everything is interpreted in terms of the participants' own understanding of it as revealed in their talk; there are no appeals to wider social rules or to extraneous contextual factors.

The relationship between context and talk is, of course, fundamental and of immediate relevance to a study of groups operating within an organisational environment. The problem is that, as Goodwin and Duranti note, a precise definition of context has so far eluded researchers, in part because of its 'dynamic mutability' (1992: 5). An immediate challenge is the potentially infinite regress that faces any analyst seeking a fully contextualised account: there is no a priori basis for saying that the limits of a particular context have been reached and explanatory resources are at an end. It is this lack of precision in determining what can be invoked as legitimate context for explanatory purposes that accounts for CA's insistence on procedural relevancy, which means that it is not sufficient merely to show that the context is relevant for the parties, it is also necessary to demonstrate how the context is consequential to the parties' conduct in a specific context (Schegloff, 1991b). As we saw when considering Schegloff's discussion of a hospital context, the conversation analyst argues that any *relevant* orientation to a specific context will be detectable in the talk itself as an accountable aspect of the participants' own construction of their shared understanding, thus rendering redundant and potentially misleading any attempt to import contextual detail from outside the talk itself.

An orientation to the process of production also foregrounds the reflexive relationship between talk and the context in which it is produced, for although the talk may be in part defined by a particular context it also serves to define that context. Every time doctors ask, 'What can I do for you?' they are not only responding to a particular state of affairs but also confirming that state of affairs through their action (for a fascinating study of the important difference between this question and 'How are you?' in the context of GP consultations, see Gafaranga and Britten, 2005). Utterances, like actions, are *context shaped* and *context renewing* (Drew and Heritage, 1993:18). A central assumption of this study is that the regular exchanges within specific professional groups will, through this process of constant renewal and reinforcement, lay down sedimentary layers of talk that, over time, give shape to a distinctive interactional landscape whose features are accessible to the analyst.

In excavating this landscape the analyst must be able to show how particular instances of talk reveal traces of sedimented understandings. 'Probably the most basic dualism inherent in individual and institution', claims Stubbs (1992: 198), 'is micro and macro', and this approach offers, if not a resolution of this profound division, at least a means of interpreting particular actions in terms of broader patterns while maintaining a connection between the two that is not possible if contextual features are simply invoked for explanatory purposes.

An example will illustrate how such a connection might be made. Extract 1.2a finds us in a staffroom during the morning break, where banter and business flow with the coffee. Students are never more than a topic shift away and Paul has just signalled a problem with one of his:

Extract 1.2a

```
001  Paul:  An:d actually (.) she's working by herself.
002         °(You know)° and- because sh- I mean she- (.)
003         she looks as if- (.) >she looks as if< she's
004         on drugs. An- and I was worried because (.)
005         sh- I said (.) she was looking up a word and I
006         said 'What are you looking for?' And she said
007         (.) 'rambling'.
```

```
008  Harry:  hhhhhhheheh
009  Paul:   t huh
010    Ed:   Fair enough. ┌Well why not.┐
011  Paul:                └And I thought┘ may:be (.) maybe
012          (.) something her landlady has said to her.
```

It is an apparently unremarkable episode, a short anecdote with a punch-line that produces a short laugh from Harry and an even shorter one from the teller, then a challenge from Ed that is brushed aside. If anything is interesting it is the way Ed's challenging assessment fails to provoke a response and his even more confrontational question, 'Well why not', never receives the reply it is designed to elicit. It is possible, of course, that the latter is lost in Paul's overlapping continuation, clearly signalled by 'And'. As an episode in itself this might provoke questions about the way anecdotes and responses to them – including challenges – are managed, or it might be treated as an example of the way students are represented in the staffroom, how specific identities are talked into being.

The situation, though, becomes much more interesting when this particular challenge is set in the wider context of staffroom talk involving this particular group. In the first place, in my 15 hours of transcribed data from this specific setting there is no other example of an overt challenge from one colleague to another, although there is an example involving an outsider. From a CA perspective facts like this are of little relevance, but if CA were to turn its attention to extended data sets from specific settings, this position might change. In this staffroom, for example, there are what might legitimately but broadly be described as challenges, but they are rare and they are managed with considerable delicacy. Not so here. In this example, Paul may not have picked up the gauntlet but it lies squarely in the middle of the floor for all to see.

Ed's response is odder yet. As ethnomethodologists have so richly shown, institutional realities, treated as objective and independent by members, are in fact created in and through those members' practices in the everyday conduct of their affairs. 'The way things are' is in fact 'The way we make things to be'. It is as well not to underestimate the power of this in our everyday affairs, or the significance of exceptions – and Ed's response is an exception. Unsurprisingly, talk about students features prominently in this staffroom, and it exhibits certain common characteristics. Students, for example, do odd things. As far as the staff-room is concerned this is an incontestable reality, the staple of back region talk and occasionally also a springboard for serious discussions of an overtly educational kind, as possible causes and responses are considered. What is never questioned is the oddness of the behaviour because for all practical purposes that's the way things are. Yet here we have Ed not only challenging Paul directly, but challenging him on the basis that this student's behaviour is *not* odd. Ed is so far out on a limb that to all intents and purposes he's shifted to a different tree. However, Paul's speculation about the landlady provides him with the chance to engage in the sort of talk that typically follows the description of an episode like this:

Extract 1.2b

```
013        Ed:   We did ⌈e:::r
014   Annette:          ⌊Well see it wasn't anything (that had
015              had come up ⌈in person.)⌉
016        Ed:              ⌊No last er  ⌋
017      Paul:   No.⌈Absolutely not.
018        Ed:      ⌊Last week we did e::m >(this you know)<
019              some work on types of er walking. That was
020              what it was. ⌈(xxxxxxxxx)
021   Annette:              ⌊O:h right.
022      Paul:   And she suddenly thought today (.)
023              'That's ⌈something I've forgotten to do.⌉
024        Ed:          ⌊That she'd look it up,          ⌋
025        Ed:   yes.
026      Paul:   Yeah. 'I could ramble.'
027   Annette:   °Yeah.°
```

Once the odd episode has been described, the teachers try to account for it, and this is what we see here, as Annette responds to Paul's speculation about the landlady with a check that the expression was not something that had come up between the student and Paul 'in person'. Meanwhile, Ed struggles to tell his story, first overlapped by Annette, aligning with Paul's position, then overlapping Annette in his turn (l.16). It is interesting to note the different responses his account prompts. Annette's 'Oh' is what Heritage (1984a) has called a 'change-of-state token' marking the receipt of new information, and the 'right' treats the information as complete. Her response, in other words, aligns with Ed's contribution as the provision of new, relevant and adequate information. Paul, on the other hand, does not treat it as adequate. In using 'and' to extend it (l.22), he draws attention to the absurdity of her decision to consult her dictionary now in order to resolve an issue that arose in a brief classroom episode a week before, with a different teacher and in an entirely unconnected lesson. Ed's overlapping completion in line 24 serves the interactional end of claiming back Paul's extension and hence associating it directly with his own account, thus effectively mitigating the absurdity of the juxtaposition. However, his 'yes' in line 25 is not emphatic but uttered with what can best be described as 'laughter in the voice', an altogether ambiguous conclusion. Paul's response to this is a reformulation of his own conclusion ('That's something I've forgotten to do.'), this time in an even more absurd form ('I could ramble.'). It is worth noting at this point that both forms of Paul's conclusion, unlike Ed's, are presented in direct speech. In this staffroom the appropriation of someone else's voice in this way is invariably deployed to humorous effect and this use of direct speech takes us back to Paul's original punchline in lines 6–7: 'And she said "rambling".'. In doing so he reinforces his representation of the student's position as distinctly odd, and Ed's challenge has now been effectively, if indirectly, dismissed.

An examination of the sequential unfolding of this episode has revealed how it is managed and yet we are still left with the nagging question of why Ed responds as he does. This is not the concern of CA, which eschews any form of psychological speculation or the attribution of motive, but the fact remains that this is a discrepant case. Every other example of talk about students conforms to the particular 'ways-in-which-we-talk-about-students' in this staffroom and, in the absence of direct evidence from the recording, we are left without any means of accounting for this single exception. Ed is a temporary teacher at the school and there are occasions when he is set apart from the other teachers, but these occasions are managed with considerable interactional subtlety, and to offer this as an account of his behaviour would be no more than speculation unsupported by any explicit textual evidence. For a solution to the problem we have to look elsewhere, in fieldnotes that record a brief exchange some months after this had taken place. Ed has now left the school and has become engaged to a former student (whom he subsequently married). Harry has just discovered this and informs Paul, who, unbeknown to Harry, has recently seen the two together and learned from someone else that their relationship began soon after the student arrived at the school. As the talk unfolds it emerges that neither of them knew anything of the relationship while Ed was at the school, even though it had lasted almost throughout that time. Neither of them mention the exchange we have just examined – they would probably be hard pushed to remember it – but from the analyst's position it explains a lot: Ed is not responding to talk about students, he is reacting to a description of his *girlfriend's* 'odd' behaviour. For obvious reasons, this is not something that Ed chooses to make relevant and his colleagues are unable to orient to what is outside their knowledge.

This is an unusual example but I think an important one because if we want to understand the workings of a particular group we sometimes have to go beyond particular interactions to the bigger picture. Arminem (2000: 443) is right to point out that it is not enough merely to identify patterns, we must also reveal their workings in order to explicate the management of institutional activities. This is why detailed analysis of particular episodes, characteristic of work in CA, is called for, and although the argument advanced in this book goes further than most conversation analysts would feel comfortable with, its attempt to connect with work in related fields builds on Schegloff's recognition of the tradition's essentially cross-disciplinary nature, located as it is, 'at a point where linguistics and sociology (and several other disciplines, anthropology and psychology among them) meet' (1991b: 46).

Our efforts, as researchers and as social beings, are directed towards an understanding of the patterns of connections that give meaning to our acts and shape to our lives, an enterprise that calls upon all our resources. The analysis that follows remains faithful to ethnomethodological principles only in so far as the categories it identifies for descriptive purposes are not invoked as explanatory resources in the analysis of discourse, which is treated as locally occasioned and participant-designed. Framed in this way the analysis is very broadly conversation analytic, but its interest in the 'underside' of organisational life is more typical of ethnographic studies (van Maanen 2001: 240-3 offers a brief discussion of issues and examples) and it draws freely on other related traditions in the spirit of open exploration that characterises much recent work in this area (see, for example, Peräkylä and Vehviläinen's appeal at the end of their 2003 paper on stocks of professional knowledge, or the way in which Stubbe et al. 2003 bring different analytical perspectives to bear on the same stretch of interaction). If the resulting analytical spread is relatively broad, this is in part because of what is needed in order to build up a sufficiently rich picture of an important aspect of institutional interaction: the extent to which talk within a collaborative professional group is designed to preserve and reinforce group identity.

A note on positioning the study

The decision to concentrate on three established professional groups in order to explore the nature of their backstage interaction precludes two potentially rewarding options: studying how a particular group develops ways of speaking, and exploring backstage encounters across a range of groups. It is therefore interesting to consider what the alternatives offer in order to better understand where the contribution of this particular study lies.

Cutting's study of the development of an 'in-group code' in a specific group (2000) is interesting because it enables her to identify features of members' talk as they change over time. She took a single group that came together for a single purpose, to complete an MSc in Applied Linguistics, collecting data from the beginning in October and noting changes as they developed between then and the end of the third term in May. The changes noted were all specific features and the outcome of this longitudinal study was a track of key changes that could then be compared, term by term. For example, the growth of informality could be detected in the large increase in the use of initial ellipsis, slang and expletives, while shared knowledge was reflected in increased implicitness and the use of contextualisation cues.

What makes Cutting's project interesting is the way she is able to identify changes over time and establish clear evidence of development. However, this achievement depends on important choices in sample selection and analytical approach that bring with them necessary limitations. For the investigation to work, the group members must be entirely new to one another and in an environment where they are free to meet as a distinct group with minimal opportunity for external participants to influence the development of the talk. In this respect Cutting's choice is inspired, but it is not one that would be viable in a business or professional context. The chances of an entirely new group being brought together in an entirely new environment would be very limited and even then it is very likely that their talk would be embedded in wider interactions within their institution. The limitation of my choice of well-established but relatively autonomous groups is that it does not allow a developmental orientation, but it does provide the basis for identifying prominent features of the group's talk, established and reinforced over time. The most important limitation of Cutting's work, though, is that while it provides a surface linguistic description, the analytical approach does not allow deeper penetration into the ways in which the talk, and hence the group's identity, is constructed. Such an approach applied to the groups in my study might have delivered an interesting description of their language but could have said nothing about what they *do*, and it is only by understanding this that we can say anything interesting about wider professional issues.

Holmes and Stubbe's work (2003) does allow interesting things to be said about the professional world, how it is constructed and the forces acting within it. The authors draw on a large database of around 2000 interactions of between 20 seconds and two hours, involving 420 participants in 14 different workplaces ranging from small businesses to government organisations. Their data collection methods embrace interviews, recordings (sometimes including video recordings) and participant observation. Although this produces an extremely rich and varied database that has over the years yielded a number of stimulating papers, many referred to in this book, such a varied collection brings its own challenges and concomitant limitations. It is clearly impossible, for example, to incorporate a wide range of extracts and at the same time dedicate the space to each that will allow fine-grained analysis. The authors' solution is to focus on power and politeness, thus exploiting the richness of their data while concentrating their analytical efforts. Interestingly, they acknowledge that many of the workplace groups studied could be described as communities of practice (Wenger, 1998), but while they recognise the significance of this their focus is not on groups as such. My own study covers some of the same topics as Holmes and Stubbe (e.g. humour, meetings) but its concentration on specific groups allows a deeper and broader analytical range within a much narrower sample.

Some of the limitations of the present study emerge clearly from a consideration of these two earlier publications, in particular its lack of a developmental perspective and its more concentrated sample. However, unlike the two earlier studies, it does offer an opportunity to explore the interactional and existential fabric of collaborative groups in a way that reveals aspects of its finer patterning. In this respect, it is closer to the work of Holmes and Stubbe than that of Cutting, and might even be seen as complementary to such larger studies.

The settings

The three settings from which data in this book are drawn cover three different but related professions with certain important features in common. One is a small language school in a market town (teachers), another a unit within a university (academics) and the third a geographically independent research unit within a larger organisation (researchers). These are small groups of professional people who have worked together for a reasonably long time (in the case of the school for 15–18 years), in a team that they would describe as collaborative and supportive, and in the face of external pressure arising from the fact that they are expected to be self-financing. These are people who are expected to deliver and do, all of them receiving very favourable external reports. Although all three groups have considerable autonomy in decision-making and are essentially non-hierarchical, each has a director/principal appointed by, and answerable to, the parent organisation; all feel confident in their professional skills but none is optimistic enough to believe that these will be sufficient to guarantee its survival in the future. Each group is, of course, different and each will have its own ways of doing and being, but there is sufficient similarity here to make the prospect of looking at all three a tempting one.

The only characteristic that requires further elaboration is that of collaboration, where self-description may not constitute adequate warrant in itself. The interactional evidence for this provided in the chapters that follow supports the description, but a previous study on collaborative culture (Nias et al. 1989) identified key criteria, all of which can be applied to the groups in this study, with the possible exception of the last if this is interpreted in institutional rather than group terms:

- a sense of independence and collective responsibility
- recognition of the need for a high degree of occupational competence
- hardworking staff with professional pride
- a sensitive and informal head
- person-centred talk
- the staffroom as 'hub'
- humour
- the selection of staff who share the institution's existing values

The following brief descriptions provide a flavour of the relevant settings.

1. The school (Pen)

Finding themselves unexpectedly redundant, the Pen teachers were given the unusual opportunity by Kate, the owner of a language school in nearby Inkham, to establish their own school and, in choosing a building and adapting it as they did, they aimed to create the atmosphere of a home rather than a school. To a large extent they succeeded and an external auditors' report makes a particular point of stressing the warm and friendly atmosphere of the school. It is located on a main road, its Georgian front consistent with the houses and offices making up the immediate neighbourhood. With the exception of a sign to the left of the door, there is little to indicate that this is not a substantial town residence: downstairs windows are curtained, upstairs windows have blinds and the front door, with its bell in the middle, is typical of many in the town. The illusion of domesticity persists at least as far as the hall, beyond which elements of institutionality creep into the furnishings, albeit relieved by softer touches. Teaching rooms are small, reflecting the emphasis on small classes and reinforcing the informal feel of the school. There are two on the ground floor, located to the left of the corridor past the staffroom, two on the first floor and a further two on the top floor.

The Pen has a core staff of five who have been teaching together for between 15 and 17 years. Harry and Paul teach general English, which tends to attract younger and more academically or socially oriented learners, while Annette and Louise concentrate on business English and attract corporate clients, many on a one-to-one basis. As principal, Jenny deals with both sets of clients. Joining this core group are temporary teachers, many at the start of their careers. Most stay between three and six months, moving on to more permanent posts, often overseas, but at least one teacher returns every summer. The post that raised most problems during my stay was that of social organiser, attached to the offer of a job at the school and occupied, with varying degrees of success, by different temporary teachers. In the case of one, Ed, it was a source of particular dissatisfaction, but most seemed to accept it as part of their deal with the school. Supporting the social programme as well as all other activities in the school is Pat, the administrator/bursar, who occupies the front office.

I spent a total of 12 working weeks as a participant observer in the school over a 15-month period, collecting data in the form of fieldnotes, interviews and tape recordings of breaktime talk and staff meetings. For the first nine months, while gaining the confidence of the teachers, I relied on fieldnotes alone, taping staff meetings towards the end of the first year and a preparation day at the beginning of the second. Throughout the first (10 week) term of the second year I taped one morning break (20 minutes) each week, choosing this particular break because it was the one that brought all members of staff together and was always the most animated.

2. The university (IEC)

There is nothing of the domestic about the International Education Centre (IEC): perched at the top of an anachronistic brick-built block that declares its dated institutionality to visitors and passers-by alike. A modern atrium and external sky lifts successfully create a delusively congenial environment for visitors to the university, but this is no more than a temporary respite for those on their way to the Centre. Many universities in the UK seemed to be designed as initiative tests for prospective applicants and this confusion of lifts, floors and wings is no exception. The Centre is located on a floor which, without rising or falling, manages to move from the tenth to the seventh and back up to the ninth floor via a serpentine corridor that almost, but not quite, completes a circuit around the buildings. Lost souls are a fact of life here.

The Centre is strung out along a meandering corridor, the stretch it occupies best described as an elongated 'S' with a double kink on the downstroke. Visitors standing in front of the double fire door with the IEC sign above it face a twisting journey involving 11 turns and five sets of fire doors before they emerge from the far end of IEC territory, except that this is not, strictly speaking, all the Centre's territory – it just happens to be where it is distributed along with other elements of the university, safe from all but the most determined conflagration.

At the end of the period of data gathering the Centre had six full-time academic members of staff (Sophie, Jack, Liz, Paul, Pip and Tony) and four part-time colleagues (Helen, Jill, Lisa and Tara), supported administratively by Anne, Kathy, Julie and, on a part-time basis, Sally. This produces a full complement of 14, though retirements mean that James and Mark also appear in earlier exchanges. There are two distinct groupings within the Centre, the first comprising Liz and Pip, supported by Anne, who are responsible for the short courses (pre-MSc teacher training, communication skills, etc.). Everyone else is involved in work on the Master's programme, and Sophie, Jack and Tony also have extensive responsibilities in the research area, especially with regard to PhD students. Julie is responsible for financial administration of the Centre and Jack is the Director.

In the Pen school I was a participant observer, but here I am an insider, a full member of the group being studied, which necessarily limits what is useable as legitimate data. I therefore used neither observation nor interviews as data sources. Audio recordings, made over a period of approximately three years, were mainly of staff meetings (each lasting approximately one hour) but also included office talk and a half-day moderation meeting involving representatives from another organisation. Most of the staff meetings were recorded by a colleague working in an area similar to mine and in these the microphone is in a central position visible to all. Permission for the office recordings (which specified the relevant conditions for these) was obtained nearly a year before they were made with a hidden recorder. Where my analysis draws on my insider's knowledge, this relates only to information available to all of the participants in the relevant meeting.

3. The research unit (DOTS)

DOTS is a fairly typical example of a small-scale rural development not too far from the outskirts of a town. It is a new but unprepossessing collection of buildings and outbuildings with a concrete forecourt and access to surrounding fields. Outbuildings apart, it is essentially a rectangle about twice as long as it is wide, with the forecourt taking up something less than a quarter of the space and the remaining areas reflecting the three broad areas of activity in the Unit: office work, laboratory work and animal work.

The staff organisation within DOTS reflects the nature of its work and the tasks associated with this. The one principal research scientist (Julie) and three senior research scientists (Cameron, Margaret and Joe) are all educated to doctoral level and are occupied with research and consultancy activities, which means that they are responsible for the design, monitoring and writing up of investigations relevant to these. This in turn means that they need to prepare research bids, liaise with clients, oversee the development of projects and prepare the reports and presentations that are their outcomes. The daily routines on which these depend are the responsibility of the scientific officers and technicians, the latter (Tina and Tim) mainly responsible for animal care and the former (Jane, Tom and Paolo) more directly involved with basic scientific procedures such as filtering, weighing and data entry. The key link between these two groups is a pair of senior scientific officers, Sue and Penny, who are responsible for drawing up weekly plans and ensuring that these are implemented successfully. This demands a grasp of a project as a whole and of its research methodology in order to discuss this where necessary with the research scientists, while at the same time they must be familiar with the wide range of practical issues that bear on basic research procedures if they are to be able to engage effectively with technicians and scientific officers.

The Head of the Unit, John, is supported by two part-time administrators, Anne and Fiona, who also maintain the front office and are responsible for the general administration of the Unit. Their work may involve them in activities as diverse as organising visits and seminars (in one case, a full-blown conference) as well as occasionally supporting other research scientists, including Julie, John's deputy. The Unit also regularly hosts a visiting overseas research student who contributes to its activities while pursuing doctoral studies.

Although this group is properly described as a team and although it shares an explicitly collaborative orientation in which all voices have the right to be heard, there are important differences between its professional structure and that of the other two groups. Superficially, all three groups are made up of colleagues in different categories of employment, the Pen comprising a principal and full- and part-time staff responsible for either business or general English, the IEC made up of academic, academic-related and non-academic staff at different levels responsible for professional or academic programmes, and DOTS divided as described above. However, in the case of the Pen and the IEC any hierarchy that the institution might recognise is set aside in day-to-day business so that these two teams work together as though all are employed on the same basis, forming sub-groups according to the needs of particular projects rather than according to a pre-set structure. This does not apply in DOTS because of its recent conversion from a civil service department to an independent company as the result of a management buyout.

The civil service is a hierarchical organisation with a distinct structure and all that is associated with this in terms specific responsibilities, lines of reporting, management, appraisal, etc. This, and the nature of the work involved, encourages much greater segmentation than is the case in the other two settings. Although there is a sense in DOTS that each section of the group has its part to play in the teamwork that produces successful projects, close collaboration tends to take place within specific teams at different levels and people tend to speak to their particular areas of expertise rather than to the more generally distributed topics characteristic of the other two settings.

My exposure to the field here was much briefer than at the Pen, amounting to only one week of non-participant observation during which detailed fieldnotes were taken. This was supported by more than 15 hours of recordings of a range of meetings as well as common room and office talk made over a period of about two years, and interviews with staff over a similar period.

Outline of the book

The shattering impact of an evaluative hammer on the fragile constructions of interactional analysis suggests *sotto voce* delivery of the claim at the heart of this book: that the considerable interactional energy these groups expend in the construction of collaborative group identity may pay distinct professional dividends, but this comes at the cost of loss of sensitivity to external contingencies. Most of the book is taken up with establishing how the centripetal dynamic shapes interaction in a range of talk types, while the final two chapters indicate how the forces that pull the group together serve also to separate it from external influences.

Chapter 2 completes the present introduction to relevant interactional issues, using a single extract to illustrate how the group dynamic works and how this relates to studies of the behaviour of small groups, in particular those with an institutional orientation. The next chapter takes potentially the greatest challenge to group unity, disagreement, and shows how this is used to reinforce group solidarity, a process that calls for some revision of standard models of argument. Chapters 4 and 5 build on this idea more conventionally, examining how humour and stories function as shared resources available to group members, serving as structural resources for the reinforcement of collaborative identity. The establishment of common perspectives is also the theme of Chapter 6, which introduces a range of other interactional resources exploited by the group to reinforce their own identity, while Chapter 7 examines ways in which 'the other' is invoked and managed within the discourse of the group, showing how talk can be constructed to displace problems and deflect responsibility for them from the group onto outside agencies. The book concludes with a return to issues of identity and interaction, indicating the contributions this sort of research can make to a better understanding of workplace communication and relationships.

If what emerges from this is typical of collaborative interaction in organisational contexts, it would seem that a considerable amount of interactional and professional energy is invested in group maintenance. It is a commonplace that marriage is something you have to work at if it is going to succeed, and friendships, as Johnson memorably observed, have to be kept in good repair. The same goes for groups: a good collaborative group will be productive, but as in all things there is a price to be paid.

2
Collaborative Identity: The Power of the Group

> We can then say that one of the most interesting things about evolution is the way in which the internal coherence of a group of living beings compensates for a particular perturbation.
>
> (Maturana and Varela, 1998: 118)

As relationships evolve – new members join, and external conditions change – the internal coherence of a group will indeed subtly adjust to small disturbances to its normal patterns of behaviour, but what is remarkable about well-established collaborative groups is the extent to which they are able to resist challenges to their way of doing things. The extract on which this chapter is based reveals the subtle ways in which such a group can flex its interactional muscles to ensure that normal routines are observed when one of the group's longest established members seeks – albeit innocently – to subvert accepted procedures. Her attempt to act as temporary Chair of one of the group's regular weekly meetings is starved of interactional oxygen from the outset, yet it withers without even so much as a hint that her actions are unacceptable.

'Power', as Arendt notes, 'is never the property of an individual; it belongs to a group' (1970: 44) and it is at its strongest when least overtly exercised. The gradually unfolding expression of the group's concerted power will be analysed in detail immediately after the extract and will provide a point of departure for the consideration of broader issues, addressing first the nature of meetings in business (with a contrastive example from the same meeting), then aspects of groups and group identity. My aim here will be to make connections with work in social psychology in order to develop the argument that micro-analysis of the sort featuring in this chapter can make not just a useful but a necessary contribution to this field.

Exhibit

Background

Each Tuesday IEC colleagues hold a staff meeting lasting from 1pm to 2pm. The meeting, which usually involves around 10 people, is chaired by the Director of the IEC and follows an agenda, copies of which are provided to participants at the start of the meeting. This, together with the fact that minutes are recorded and forwarded to a university committee as part of the flow of information within the institution, means that the meeting is technically formal rather than informal. The circumstances today, however, are slightly unusual because the Chair is late. At his urging there has been a concerted effort recently to ensure that all meetings start promptly, a circumstance which is possibly unique in the university, where even the most formal meetings seem to allow for a period of about five minutes' grace. Today, however, Jack is teaching. So far, those teaching in the hour up to 1pm have arrived about five minutes into the meeting and there has been a tacit understanding that this is acceptable, but this is the first time that Jack, the Chair, has been in this position and the option of beginning at 1pm without him has not been discussed. In the past, whenever he has known that he would be unable to attend a meeting, he has nominated another colleague to act as Chair and there is no precedent for self-election to this role.

At 1pm Jack has still not arrived and the nine people present (all academics, with the exception of Sally and Kathy, who are administrators) are talking among themselves. There are at least three separate conversations taking place at the same time, a fairly typical situation prior to the formal opening of the meeting. Two of these conversations merge into a single exchange on an administrative issue and the third gradually trails off, eventually leaving only one conversation with Tony and Helen as the main participants. Interestingly, once talk has shrunk to a single conversation, this remains the pattern and the speakers, although essentially engaged in a private exchange, do so in the presence of 'overhearers'. This perhaps explains the occasional long silences that occur, and it is at the end of one of these silences, lasting eight seconds, that Helen decides to move things forward. The clock on the wall shows almost five past one, there is still no sign of Jack and the agenda is a full one, so her decision is understandable. Its effects, however, are not as she might have predicted.

Text

Extract 2.1

```
001    Helen:   Can we start for instance.=
002    Kathy:   =m
003    Helen:   Can anyone think of a way of getting Paul Klein (.)
004             a scholarship
005             (0.2)
006    Helen:   to finish his Master's.
007             (0.8)
008    Helen:   And do you all think he's a deserving enough case.
009             (4.5)
010     Tony:   For whom?
011    Helen:   Jan
012             (0.2)
013    Helen:   Klei::n.
014             (0.8)
015    Helen:   >Oh no wait a minute it< wasn't Jan Klei:n was it?
016             (2.0)
017     Paul:   Jan Klaster.
018    Helen:   It's not Jan Klaster, it's H: a ⌈ns.
019     Paul:                                  ⌊Hans Klaster.
020             (0.4)
021     Paul:   (Cc(x)⌈c(x)ccee)
022    Helen:         ⌊Hans ⌈KlO:ster.
023    Sally:              ⌊Kh!haHH!hehhehehe ⌈he
024    Kathy:                                ⌊HOhoho-!
025    Helen:   No=W:onder everyone's looking at me blank.
026             (1.0)
027    Helen:   >Well< it's Hans Kloster.
028             (3.5)
029    Helen:   Okay.
030             (3.0)
031    Helen:   He was on campus.
032     Paul:   I did mention ⌈this to someone ⌉ (.) the other day,
033    Helen:                 ⌊A few years ago.⌋
034             (0.5)
035     Paul:   ⌈⌈And we thought it wasn't (alright)⌉
036    Helen:   ⌊⌊And (.) he's still- >I mean< he's ⌋ (.)
037             °bloo:°dy good assignments. Rea:lly really good
038             ones. Working (.) out of China:, working in: >the<
039             Chinese::s:ystem and everything.
040    James:   ((Coughs))
041    Helen:   I'm just wondering if we can think of a source of a
042             bursary for him to finish it.
043             (0.5)
```

```
044  Kathy:  Mmm
045          (0.5)
046  Helen:  Because it would be good wouldn't it?
047          (0.4)
048  Helen:  If y- if we could get him through the Masters.
049          (1.5)
050   Tony:  This is wonderful. ((Addressed almost to self, not
051          referring to topic.))
052  Helen:  (China::)
053          (1.0)
054  Helen:  MMmm!
055   Tony:  Ss- I'm sorry.
056  Helen:  No ⌈it's all right.
057   Tony:     ⌊No (xxxxxxxxx).
058   Tony:  (It's it's) a wonderful bit of data I repeat it
059          second hand I suppose.
060          (0.5)
061   Tony:  I said to Eleni (.) >when I saw her in the<
062          corridor,
063          (0.5)
064   Tony:  'What have you done with Jack?'
065    ??:  Gh ⌈m!
066   Paul:     ⌊Mm ⌈Mm
067    ??:          ⌊HHhhh!=
068   Tony:  =Answer,
069          (1.0)
070   Tony:  'cross cultural communication'
071          ((General laughter))
072          ((Jack returns during the laughter, explains his
073          absence and starts the meeting.))
```

Analysis

An unsuccessful opening move

Extract 2.1a

```
001  Helen:  Can we start for instance.=
002  Kathy:  =m
003  Helen:  Can anyone think of a way of getting Paul Klein (.)
004          a scholarship
005          (0.2)
006  Helen:  to finish his Master's.
007          (0.8)
```

```
008  Helen:  And do you all think he's a deserving enough case.
009          (4.5)
```

Helen's opening move is a rather surprising one, emerging unilaterally, making no concession to the topic currently under discussion, and offering no opportunity for the group to negotiate a response to her proposal. It is also surprising because of its intonational contour: the falling profile at the end of her first turn suggests that 'for instance' is connected to starting, but it could also be read as attaching to her second turn as an 'instance' of an agenda item, plunging straight into business.

It is understood that prior to meetings those present engage in conversation, and although the subjects covered tend to relate to work, speakers enjoy the usual freedoms applying to conversation: anyone can contribute, there are no predetermined topics and all speakers have equal rights. Up to this point Helen has oriented to these norms, but her proposal to start the meeting represents a significant shift to a new activity with very different interactional rules. The fact that her proposal is formulated syntactically as a question has to be set against its intonational contour, the ambiguous placement of 'for instance' and the immediate introduction of an agenda item. Other participants, still engaged in conversation, have no time to respond to Helen's initial proposal, which effectively turns out to be not the request, 'Can we start' but a signal: 'We are starting.' Kathy's minimal response, latched to the end of Helen's turn, constitutes a response of sorts, but its significance in this context has to be interpreted in the light of the fact that Kathy, secretary to the programme, is effectively an observer who rarely participates in discussions unless explicitly invited.

It is up to those present to decide how they will respond to Helen's action, but Wilson provides a memorable example of what the interactional consequences of such topic nomination in ordinary conversation, albeit among children, can be:

Extract 2.2

```
001  L.  I know we'll talk about all these troubles
002      and our solutions to them.
003  D.  We'll do no such fucking thing
004  P.  Aye just cause your da got shot
005  D.  [an my ma got killed (ah ahahaha)
006  P.  And your dog died (laughter)
```
                                          (Wilson, 1989: 25)

Of course, the institutional setting in our example to some extent legitimises Helen's move and makes any such outcome unlikely in the extreme. In any case, generally speaking adults are able to find less colourful ways of making interactional points, as we shall shortly see. First, though, consider another interesting feature of Helen's turn. In raising a specific topic (1.3) she has done more than simply attempt to shift to a new activity, for whenever we speak we also reveal something of ourselves, our understanding of what is going on and how we relate to those we are addressing. Here, by beginning the meeting and nominating the topic, Helen is effectively establishing herself as Chair, albeit temporarily given that we can expect Jack eventually to appear and take over this role. As Chair she has the right to nominate the topic and select speakers, so in making this contribution Helen has bitten off a fairly meaty chunk of responsibility. In fact she has done more, although the transcript doesn't show this. Everyone present has the agenda for this meeting and it is standard practice for the Chair to signal any deviations from this agenda – the data offer plenty of examples of this, always involving an explanation for the change – but Helen has plunged into what happens to be the fourth item on the agenda without preface or explanation. The fact that this particular item has her name next to it gives her, to some extent at least, proprietorial rights over it, and the use of 'for instance' to some extent mitigates the force of her imposition. Nevertheless, there is still no precedent for an action of this kind.

Also interesting is Helen's shift from 'we' to 'you'. Her opening is an inclusive appeal to the group, constructing herself as part of the collective self, 'the "we"' as 'opposed to the "I"' that 'lies at the heart of the perceptual, attitudinal, and behavioural effects of group membership' (van Knippenberg and Ellemers, 2003: 31). But once the topic is introduced, she switches to 'you', establishing herself in a dialogic relationship with the rest of the group and implicitly underlining her assumption of a distinct identity as Chair. The significance of personal pronouns for identity work in meetings has been noted (e.g. Fasulo and Zucchermaglio, 2002) and Iedema and Scheeres (2003: 323–4) identify in their data a similar shift between 'we' and 'you', where the speaker involved constructs herself first as part of and then separate from workers on the factory floor.

The response of the group to Helen's move is profoundly more subtle and more bruising than the individual response of D in Extract 2.2 above. It depends on the exploitation of a particularly powerful interactional feature known as the *adjacency pair* (Schegloff and Sacks,

1973: 295–6). Such pairs consist of two parts: a first pair part designed to elicit a response and a second pair part designed to provide such a response. On completion of the first pair part the first speaker stops and the next speaker is expected to provide the second element. Examples are very common and although they take various forms such as question and answer, greeting and greeting, invitation and acceptance/refusal, they all share the same essential quality: the second pair part is predicted by the first and its absence is significant. The absence of a response is more than mere silence, it is an 'accountable action' (Sacks, 1992a: 4). Whether or not anyone actually provides an 'account' of the behaviour is beside the point; it is treated as in some sense motivated. Levinson (1983) calls silences of this sort 'attributable' to distinguish them from mere gaps or lapses in conversation for which participants cannot be held to account.

When in lines 3–6 Helen provides a first pair part inviting colleagues to think of ways of obtaining a scholarship for a current student, there is a silence of just under a second, prompting her to ask a further question which opens up the possibility that those present might not think Jan Klein a sufficiently deserving candidate for such a scholarship. This extension is, in fact, a form of accounting for the prior silence because if those present do not think that Jan meets this criterion, they have no basis for responding to her first question. Her second question directly addresses the others around the table, this time appealing to 'all' of them where before she had asked for a response from 'anyone', but it proves no more successful than the first and is followed by a silence of considerable duration – over four seconds. The reason for the silence does not become immediately apparent.

Clearing up a misunderstanding

Extract 2.1b

```
010    Tony:  For whom?
011    Helen: Jan
012           (0.2)
013    Helen: Klei::n.
014           (0.8)
015    Helen: >Oh no wait a minute it< wasn't Jan Klei:n was it?
016           (2.0)
017    Paul:  Jan Klaster.
```

```
018   Helen:   It's not Jan Klaster, it's H:a┌ns.
019   Paul:                                   └Hans Klaster.
020            (0.4)
021   Paul:    (Cc(x) ┌c(x)ccee)
022   Helen:          └Hans ┌KlO:ster.
023   Sally:                └Kh!haHH!hehhehehe ┌he
024   Kathy:                                   └HOhoho-!
025   Helen:   No=W:onder everyone's looking at me blank.
```

Since conversation is jointly constructed by the participants involved, there is always the possibility that something will go wrong; perhaps a speaker begins a word and then decides that another would be preferable, a hearer fails to hear properly or misunderstands what has been said, or someone offers an inaccurate version of events. Anything that might cause 'trouble' in the conversation, however minor such trouble may be, is a candidate for repair by the participants so that the conversation can be brought back on track. Conversation has been described in terms of architectural design (see Heritage, 1984b: 254–60 for a discussion of the *architecture of intersubjectivity*) and the building of shared understanding, and repair can perhaps best be thought of in terms of putting right something that has gone wrong in the process of construction. It may simply be a matter of correcting a minor fault, but sometimes, as here, it involves careful reconstruction by the participants in order to ensure that building can continue.

In this case the *trouble source* is in line 3 where Helen names the student. Because Tony has either not heard the name or can't fit the person named into the context of the need for a scholarship, he initiates a repair sequence that will eventually produce the correct name (1.22). At first Helen repeats the original name, but after a short pause recognises that this is not the correct name and appeals for help in pinning it down (1.15). Eventually, a collaborative effort by Paul and Helen produces the correct name and completes the repair sequence, to the amusement of Sally and Kathy. Repair trajectories like this one are often no more than a *side sequence* (Jefferson, 1972) to the main topic, which resumes as soon as the trouble has been resolved, but in this case the confusion over names provides Helen with an explanation for the 'blank' looks which followed her question. Now that the misunderstanding has been resolved, she invites a response by repeating the correct name – with similarly unfortunate results.

Pressing on regardless...

Extract 2.1c

```
026          (1.0)
027   Helen:  >Well< it's Hans Kloster.
028          (3.5)
029   Helen:  Okay.
030          (3.0)
031   Helen:  He was on campus.
032    Paul:  I did mention ┌this to someone ┐ (.) the other day,
033   Helen:               └A few years ago. ┘
034          (0.5)
035    Paul:  ┌┌And we thought it wasn't (alright) ┐
036   Helen:  └└And (.) he's still- >I mean< he's ┘ (.)
037          °bloo:° dy good assignments. Rea:lly really good
038          ones. Working (.) out of China:, working in: >the<
039          Chinese::s:ystem and everything.
040   James : ((Coughs))
041   Helen:  I'm just wondering if we can think of a source of a
042          bursary for him to finish it.
043          (0.5)
044   Kathy : Mmm
045          (0.5)
046   Helen:  Because it would be good wouldn't it?
047          (0.4)
048   Helen:  If y- if we could get him through the Masters.
049          (1.5)
```

As Pomerantz notes in her examination of pursuing a response (1984b: 161), '[i]f a speaker expects a recipient's support or agreement and instead the recipient displays difficulty in responding, the speaker would be motivated to figure out what went wrong and to remedy it.' To all intents and purposes, Helen would seem to have achieved this, and yet the resolution of the apparent misunderstanding produces no change in the group's response; in fact, the silences in lines 26 and 28 almost echo the earlier pair of silences in lines 7 and 9. Treating this as an indication that more information is required, Helen identifies Hans, who is now a distance learning student, as someone who once studied on-campus. This prompts a negative reply from Paul (ll.32 and 35), but his message is either lost in the overlap with Helen (ll.35–6) or deliberately ignored by her as she goes on to offer an explanation of why he deserves a scholarship. Her description of Hans's performance is designed to generate the positive response that has so far eluded her, but when it proves no more successful than her earlier efforts she reformulates her original 'Can anyone think of a way...' (l.3) much more indirectly as something she is 'wondering' about. Significantly, she also positions herself as one of the group, shifting from her original 'And do you all think' (l.8) to 'if we can think'. Despite these efforts, the only contribution from the group is a cough and an acknowledgement from Kathy, who as an administrator would not be involved in the decision anyway. After yet another silence, Tony speaks, but his contribution will serve only to draw a line under Helen's unavailing efforts to start the meeting.

... and getting nowhere

Extract 2.1d

```
050    Tony:   This is wonderful. ((Addressed almost to
051            self, not referring to topic.))
052   Helen:   (China::)
053            (1.0)
054   Helen:   MMmm!
055    Tony:   Ss- I'm sorry.
056   Helen:   No ⌜it's all right.
057    Tony:      ⌊No (xxxxxxxx).
058    Tony:   (It's it's) a wonderful bit of data I repeat
059            it second hand I suppose.
060            (0.5)
061    Tony:   I said to Eleni (.) >when I saw her in the<
062            corridor,
063            (0.5)
064    Tony:   'What have you done with Jack?'
065     ??:   Gh ⌜m!
066    Paul:      ⌊Mm ⌜Mm
067     ??:          ⌊HHhhh!=
068    Tony:   =Answer,
069            (1.0)
070    Tony:   'cross cultural communication'
071            ((General laughter))
```

Tony's contribution here has to be interpreted in the light of Helen's original move, which cut across their 'public' conversation and attempted to move the talk onto a formal agenda. So far nobody has given any indication that they have accepted this move, and in this sense Tony's contribution is entirely legitimate. However, because it represents a rejection of Jane's bid it calls for careful and subtle management. We can only speculate on whether Tony deliberately introduces it and merely frames it as an example of thinking aloud, or whether it is genuinely accidental; all that matters is that although it serves as a *story preface* (Sacks, 1992b: 10) and therefore represents a bid for an extended turn (ibid. 225-6), it also emerges as unintended.

Accidental interruptions of this sort usually call for an apology, and one is immediately forthcoming. Helen accepts this, but there is no extended and delicate negotiation of face by Tony (l.55), just a slight hitch in his reformulation of the preface, a brief pause then the story itself, one which fulfils all the main conditions for a successful anecdote: a relevant topic (Jack's absence), freshness (it occurred just prior to Tony's arrival) and humour. The move from putative agenda back into pre-meeting talk is underlined by the general laughter that follows the punchline, as eloquent a statement of the group's position as its earlier silences.

Discussion

Introduction

In the context of our daily business, identity is not the essence of personhood, an identifiable and ineradicable pith to which we attach our other attributes, it is an interactionally constructed representation that serves our social needs. If I am not the same person with my family at the dinner table as I am when I speak at a conference, this is as it should be and not the result of some deceitful charade on my part. In constructing a different identity in these two situations I am responding to a complex of social and interactional expectations that bind me and other participants in a functional relationship that makes it possible to achieve personal and social ends. Announcing my identity will not serve as a substitute for this – it must be sanctioned by the interactional behaviour of those present. As this exhibit shows, when we attempt to establish an identity as part of a personal agenda, there is always the possibility that this will be successfully resisted by other participants.

In order to appreciate more fully the intricate ways in which the talk here is managed, it is necessary to understand something of how participants are able to orient to an activity and maintain that activity as part of 'normal business'. The ability to do this depends, more broadly, on a shared understanding of 'what it is we do' and 'how we do it', rather than on any explicitly stated rules or rituals, and the interactional evidence is itself reflexive: because talk is constituent of the activity in which the participants are engaged it offers the means by which they as insiders can define the activity. And what is understood by the participants in this case is the activity of 'being engaged in a meeting'. The fundamental importance of this understanding will be starkly revealed through a comparison of how the same agenda item is dealt with inside and outside the 'meeting'.

The way we do things here

The most powerful explanation of the pervasive strength of shared ways of doing things and its capacity to dominate our understanding is to be found in the tradition of ethnomethodology, mentioned briefly in Chapter 1. The term was coined by Garfinkel (1974) as a means of capturing his surprising discovery that the deliberations of a jury arose not from any attempt to engage in quasi-legal reasoning but from the application of an ordinary or 'folk' (ethno-) approach to decision-making (methodology). The subject matter of this tradition, simply put, is the study of common sense and its workings, its primary activity an exploration of the complex and yet – to participants – blindingly obvious way in which ordinary people make sense of their world and deal with unfolding circumstances.

Central to the ethnomethodological account, and a key to understanding what is happening in the IEC meeting, is its view of social norms. One way of interpreting social behaviour is to see it as regulated by such norms, but ethnomethodology offers a subtler and more profound account of our actions:

> ...the common norms, rather than regulating conduct in pre-ordained scenes of action, *are instead reflexively constitutive of the activities and unfolding circumstances to which they are applied.*
> (Heritage, 1984b: 109. Italics in original.)

Every time I walk into the general office and greet my colleagues there with a cheery good morning, the enactment of this greeting is not only a reflection of our understanding that this is normal, it is also a reinforcement of that normality. This may seem obvious enough but the power of such reflexivity should not be underestimated since it implies that the absence of an expected response is more than a failure to act conventionally: it constitutes a different activity and potentially a different agenda. Implicit in the ordinariness of our everyday behaviour is the assumption that it is explicable as simply common sense and not susceptible to further explanation.

Garfinkel's work revealed a wonderfully complex set of social practices and associated understandings arising from the capacity of individuals to develop procedures for making normal the situations they encounter. He exposed the intricate and involving beauties of a social world hidden behind the modest representations of 'common sense', whose particularities are paradoxically at their most impenetrable to those most intimately familiar with them. Nowhere is this more obviously demonstrated than in the assumptions underlying our everyday exchanges:

> The anticipation that persons *will* understand, the occasionality of expressions, the specific vagueness of references, the retrospective–prospective sense of a present occurrence, waiting for something later in order to see what was meant before, are sanctioned properties of common discourse. They furnish a background of seen but unnoticed features of common discourse whereby actual utterances are recognized as events of common, reasonable, understandable, plain talk. Persons require these properties of discourse as conditions under which they are themselves entitled and entitle others to claim that they know what they are talking about, and that what they are saying is understandable and ought to be understood. In short, their seen but unnoticed presence is used to entitle persons to conduct their common conversational affairs without interference. Departures from such usages call forth immediate attempts to restore a right state of affairs.
>
> (Garfinkel, 1967: 41–2)

The argument in this book is founded on the conventional claim that all groups develop characteristic ways of speaking and acting, and nowhere have the implications of this been more subtly explored than in the work of the ethnomethodologists. Although their rigorous methods, which demand much of the researcher in terms of time, skills and sensitivity, will not be used here, in bringing out 'seen-but-unnoticed' aspects of the interactions of the three collaborative groups, the analysis implicitly acknowledges their insights. The next section will consider in more detail the structure and function of meetings in IEC in order to add dimensionality to the basic picture revealed by the interaction itself, but already it is clear that the participants have succeeded in restoring 'a right state of affairs' in response to Helen's attempted deviation from normal practice.

Again and again, as the analysis of key exchanges unfolds, we shall see evidence of the sort of features Garfinkel identifies and the ways in which they serve to achieve the group's ends. In this extract, the simple expedient of withholding or delaying responses prevents Helen from leading the group into the meeting agenda she wishes to pursue. The group's actions are interpreted by her at first as merely a response to a mistake she has made over the name of the relevant student, but when further efforts are unavailing she does not resist Tony's anecdotal finesse into the norms of pre-meeting talk. Eventually her topic will re-emerge in the meeting proper and will be treated very differently. In order to understand why this should be so, it is first necessary to appreciate the significance of Helen's action in the context of meeting behaviour.

Minding meetings

Organisations depend on meetings, from the quickly assembled informal meeting at section level to settle a local strategy or procedure, to the formally constituted assembly of elected or nominated representatives. Few decisions, from the choice of which new coffee maker to buy to a commitment to opening a new branch in a different part of the globe, are made without being aired in one sort of meeting or another, and for this reason there is perhaps a temptation to see meetings in a purely instrumental light. However, this is to ignore their importance as social encounters through which individual, group and institutional relationships, with all their associated histories, motivations and agendas, are played out. For meetings are more than conventional means of decision making and information sharing; they are, in Boden's words (1994: 81) '*the* interaction order of management, the occasioned expression of management-in-action, the very social action through which institutions produce and reproduce themselves.' Institutional they most certainly are, but Boden's words apply equally well to the group, which exploits the meeting as an instrument of its own purposes.

As a structure carrying such complex freight, the meeting – and its organisation – represents a core activity. In fact, because meetings cover such a wide range of topics and purposes, engaging groups of such different sizes and composition, the generic term 'meeting' embraces a range of structures and activities (see, for example, van Vree, 1999). However, a distinction is commonly drawn between formal and informal meetings, usually on the basis that the former has a Chair who is responsible for ensuring that the agenda is followed. The formal meetings in all three of the settings in this study are, in fact, strikingly similar: they involve a relatively small group of people (usually from 6–12), meeting in a standard venue that is not somebody's office, with a recognised Chair and following a formal agenda. In all three settings the meetings are regular, and in the case of the IEC they take place at the same time every week during term time. The fact that something is a formal meeting may say little about its content, but over a period of at least a decade colleagues in the IEC have structured their practice in such a way that certain issues will be dealt with as part of a chaired weekly meeting beginning at a set time.

The presence of the Director, or his nominee, as Chair is a distinguishing feature of these weekly IEC meetings, and associated with this role are a number of rights and responsibilities, not least the right to open and close the meeting. In fact, it might even be said that the act of opening the meeting identifies the performer as Chair:

> The meeting can be said to commence when the predetermined Chair, or a person from the group (with its approval), initiates the opening of the proceedings. From this moment on, that individual takes on the specific roles of Chair and becomes invested with the unconditional power of opening and closing the meeting. Thus, *the openings and closings are the most rule-governed stages of meetings*: no other participant is allowed to carry them out without committing a noticeable breach of convention.
>
> (Bargiela-Chiappini and Harris, 1997: 208–9. Italics in original.)

The significance of Helen's statement 'Can we start for instance', when taken in conjunction with the immediate move to an item on the agenda, cannot have been lost on those present: the act itself identifies her as Chair and represents a clear breach of convention. Of course, the act itself is in a sense no more than a claim to this position unless it is sanctioned by those present; it is up to them whether they choose to align to her position and allow discussion of the agenda item to move forward. As we have seen, without any explicit acknowledgement of the breach that has taken place, the group manages to move back to its previous pre-meeting talk where the floor is open to all, topics are not nominated, and anecdotes and jokes are to be expected. The only response to Helen's agenda point is Paul's almost anecdotal reference to having mentioned this topic to someone previously ('I did mention this to someone (.) the other day, and we thought it wasn't (alright).'), a contribution that Helen treats almost as an aside and not one that is pursued by any of the other participants. The significance of the group's response to Helen's attempt to introduce her agenda item can be fully appreciated only if it is compared with the its reaction to the same item when it is introduced by Jack later in the meeting.

Normal business

Tony's anecdote proved to be aptly timed, allowing Jack to return on the crest of the laughter it provoked. An apology for his lateness takes the form of an extended account explaining that he had hurried from his teaching session to last week's venue in the mistaken belief that it would also be used for this meeting, a response incidentally highlighting two essential features of all meetings: time and venue.

Having arrived, he moves swiftly onto the first item of the agenda and the group soon reaches the fourth item. The most obvious difference between the response this time and the exchanges when Helen attempted to introduce it is the length of the discussion that follows, despite Jack's attempt to wrap up the item as quickly as possible:

Extract 2.3a

```
001    Jack:   Okay.
002            (1.5)
003    Jack:   Jan Klein,
004            (0.2)
005            scholarship ⌈fund.
006    Helen:             ⌊It's not Jan Klein, it's Hans Kloster.
007            I'm⌈sorry. I always get them⌉ muddled up
008    Kathy:     ⌊HuhuhuHUhuhhh             ⌋
                    ...
017    Jack:   Is⌈there a scholarship fund?=No.
018    Helen:    ⌊E:m,
019    Helen:   NE-
020            (0.5)
021    Helen:   E:m,
022    Jack:   Can they (get a scholarship?) N:o.
```

Jack's 'Okay' signals the move from the previous topic, a 'prefiguring' of movement towards next-matters (Beach, 1993: 341). After a short pause, he nominates a new one, the scholarship fund for Jan Klein (ll.3–5). Helen's overlapping repair ensures that confusion over names is not repeated, and her explanation of the mistake, together with a comment from Jack on Jan Klein, who is a current on-campus student (lines omitted), extend the sequence. However, as soon as this is concluded Jack dismisses the possibility of a fund in two rhetorical same-turn exchanges (ll.17 and 22) before Helen has had the opportunity to develop her case.

When Helen subsequently pursues an enquiry about the existence of a university fund, Jack's statement of the situation is emphatic and clearly designed to close down the discussion:

Extract 2.3b

```
044   Jack:  There's NO scholarship within the School
045          nor likely to be for anybody under any
046          °circumstances period.° And I can say that
047          >and that's not my decision< that's (.) the
048          School's decision that's Jane's decision
049          that's the uniVERsity's decision. · hhh
050          HArdship only applies as far as I understand
051          it, to on campus::,
052          (0.4)
053          students who have to have an INterview,=with
054          people from Registry, who have to apply
055          formally,=
056  Helen:  =Right.
```

The basis for his dismissal is that scholarships are not available anywhere in the university and that the only alternative source of money, from a hardship fund, is available only to on-campus students. However, the rhetorical structure of this extended turn is interesting, suggesting a pronouncement rather than an argument. It comprises three parts. The first begins with a simple statement of fact (l.44) which might have been sufficient in itself, but it is followed by a three-part emphatic expansion closing down the possibility of change through the effect of time ('nor likely to be'), person ('for anybody'), or circumstance ('under any circumstances'), rounded off by the underlining finality of 'period'. The three-part structure is picked up even more explicitly in the next part of the turn where Jack locates responsibility for the decision in the institution itself. Exploiting his own negative statement ('that's not my decision') as a point of contrast, he develops a three-part list of the form 'that's X's decision' to implicate the institution as represented by the School, the Head of School (Jane) and the University. The turn ends with a statement explaining why a hardship fund is not available, and even here there is the suggestion of the beginning of a triadic structure in the lexical and grammatical parallelism of the relative clauses ('who have to have...who have to apply') and the fact that the second ends with a continuing intonational contour, indicating that a third part is forthcoming.

Atkinson has drawn attention to the 'air of unity or completeness' that characterises three-part lists (1984: 57) and demonstrated how they are used by politicians to win applause. It is this rhetorical aspect that is apparent in Jack's turn, reinforced by the linguistic parallelism and producing an extended, almost declamatory, statement. The length of this turn marks it out immediately as different from anything that appears in Helen's attempt to open the meeting and its overtly rhetorical structure serves not only to underline Jack's case but his position as mediator between the group and the institution.

So far the exchanges on this subject have been confined to Helen and Jack, but, in marked contrast to the opening phase, the rest of the group will become involved in the discussion. Ironically, the person who brought Helen's original bid to an end is the one who moves the discussion forward by introducing a possible financial strategy that lies outside the responses Jack has dismissed:

Extract 2.3c

```
060          (0.6)
061  Tony:   We could if we wanted to
062          (0.8)
063  Tony:   could we not
064          (1.4)
065  Tony:   offer to::e:m defer payment.
066          (2.5)
067  Jack:   Don't know. In what sense? >Sorry.< Er-I
068          °I genuinely haven't understood.°
```

Tony's tentative suggestion is in marked contrast to Jack's emphatic dismissal, but even more interesting is Jack's response (ll.67–8). That it comes from Jack is entirely consistent with his role as Chair: in the intervening 2.5 seconds anyone else might have spoken, but the suggestion is effectively directed to the meeting via the Chair. Jack's request for clarification following his admission of ignorance could easily be taken as a challenge to Tony's proposal, coming as it does on the tail of the former's categorical dismissal of the case for a scholarship. Jack's apology and clarification of the illocutionary force of his question as a request for clarification closes off the possibility of such an interpretation.

What follows is an extended sequence which, in contrast to Extract 2.1, involves all of the academic participants working together to construct a shared response. The following extract gives a flavour of the exchanges:

Extract 2.3d

```
130    Jack:   We'll tell him that, but tell him that of
131            course there's noth:ing
132            (0.6)
133    Jack:   that he cannot
134            (0.4)
135    Jack:   graduate unless he's em:
136            (0.4)
137    Paul:   [[°Got the money.°]
138    Jack:   [[he's actually.  ]
139            (0.2)
140    Jack:   Y[eah.
141    Tony:    [If he didn't finally get the money
142            together and pay up within a five year
143            period, both his work and our work would be
144            down the tu[bes.
145    Jack:              [That's!=absolutely it.
146    Helen:  °mm [m°
147    Jack:       [That's absolutely it. POSSIBLY:: if
148            he's got a good enough case we could get a
149            (0.2) a WAIver for a yea:r, but at-the MOst
150            we could get it for would be for a year >and
151            it would have to be a waiver which is not a
152            straightforward process °so you couldn't
153            guarantee it.°
154    Mark:   What are his circumstances?
```

In concluding Jack's statement for him in line 137, Paul jointly constructs their representation of the position. This completion occurs just at the point where Jack continues his turn with an interpolated 'he's actually'. Jack might easily have continued to round off his statement along the same lines as Paul, but instead he adopts a stronger form of alignment: by concluding his statement with downward intonation at a grammatically inappropriate point (after 'actually'), Jack not only signals the end of his turn but allows Paul's completion to stand as the final word, with which he can then agree (l.140). The implications of the student failing to meet the relevant financial conditions are taken up by Tony and emphatically endorsed by Jack (l.145), who goes on to introduce further possible actions. Mark's contribution (l.154), on the back of this, invites further consideration. Together, then, they explore possible lines of response, closing off some and pursuing others, so that when finally a position emerges it will be one in which they are all implicated.

In fact, it emerges that Hans is no longer in China but has moved back to Holland, so there is no possibility of financial support, even along the unusual lines they have been exploring. Helen is apologetic for having failed to check this:

Extract 2.3e

```
223   Helen:  =I:yy was::
224           (0.2)
225   Helen:  jumping to the conclusion that he was still in
226           China because the last time he emailed me, just on
227           a friendly basis, was that he was still in China:=
228   Paul:   =Yeah=
229   Helen:  =an-and ┌he ┐ was really=
330   Paul:           └Bu-┘
331   Helen:  =poorly ┌paid, and that's what I thi- ┐ that's=
332   Jack:           └But I- I thought that too:, ┘
333   Helen:  =what ┌I thought it was.              ┐
334   Jack:         └and I >figured if we-< ┘if we make
335           exceptions for someone who ┌clear┐ ly mm=
336   Helen:                              └No.  ┘
337   Jack:   =we ┌didn't (xxxxxxxx that) really- (what=
338   Helen:      └>Nononononononono. No no=
339   Jack:   =the situation was.) ┐
340   Helen:  =we don't. We don't ┘we don't. Certainly ┌don't.
341   Paul:                                            └Yeah.
342   Jack:   No.
343   Helen:  No. (xxxxxxxxxxxxxxxxxxx)
344   James:  No scholarship then.
345   Helen:  No no. It's all right. (Definitely:::)=
346   Jack:   And he can't use the China one any more can he?
347   Helen:  No no ┌(no more)
348   Jack:         └Okay.
349   Helen:  I'm so ┌rry about that.┐
350   Jack:          └Jolly good.    ┘ That's all right.
```

Having begun by taking up opposing positions at the beginning of the discussion, Helen and Jack are now at pains to support one another. Jack admits to the same misconception as Helen (l.332), while Helen's responses to his conclusions builds to an agreement of rare force: 'No ... Nonononononono. No no we d<u>o</u>n't. We don't we don't. Certainly don't.'. By the time Jack signals the end of this agenda item, everyone in the team is on board and the topic itself has received a vigorous airing.

The contrast between these two extracts serves a methodological purpose as well as underlining the essential nature of the meeting as an interactional event. Methodologically, the selection of an example where a 'hitch' occurs in the predictable organisation of events, allows what Levinson (1983: 319) has described as a 'key source of verification' for the ordinary arrangement of affairs represented in the second extract. At the same time, the two events share a common feature that remains in the background in the second example but is painfully exposed in the first: although the Chair may to some extent control the floor and may nominate speakers, the turn-taking system is essentially a collaborative enterprise involving the whole group and not at the disposal of any particular individual (Larrue and Trognon, 1993: 194). To place this in its wider context a consideration of the collaborative group is called for.

The collaborative group

Although the task of pinning down a definition of groups might be likened to that of nailing jelly to a wall, the challenge has frequently been taken up, often with messy results. However, the interactional glue that binds groups together features with reassuring regularity, sometimes producing very similar positions. Bonner (1959), for example, defines a group as a number of people in interaction with one another and points to the process of interaction as the feature that distinguishes a group from a mere aggregate, a position echoed 30 years later in Forsyth's definition of a group as 'two or more interdependent individuals who influence one another through social interaction' (1990: 7). The study of a group's interaction is therefore, in a very real sense, a study of the ways in which it maintains its identity as a group.

It might, of course, be argued that because the groups in this study exist as part of institutions, the maintenance of group identity is inextricably bound up with the larger group of which it is a part. Up to a point this is true but not in any sense that subsumes the former under the latter or sees the identity of the work group as in some way deriving from the larger unit. In fact, both practically and analytically, it is essential to recognise the primacy of the work group as the locus of organisational life. Van Knippenberg and Ellemers (2003: 34) go further than this, arguing not only that work group membership, rather than organisational membership, determines day-to-day employee functioning, but also suggesting that a focus on identification with the organisation as a whole may be misplaced because behaviour induced by group membership, such as covering up for a colleague's deficiency, may work against the interests of the organisation as a whole. Commitment to the group may not, of course, be uniformly shared. In fact, research suggests that there are four types of commitment: to the work group, to top management, to both, and to neither (Turniansky and Hare, 1998: 96).

Issues of commitment, however, are difficult to pin down. More relevantly, these groups are, in Giddens's terms, communities characterised by 'high-presence availability' (1979: 103) based on constant face-to-face interaction, where a strong in-group identity has already been established. The benefits of this for the 'classic goals of communication . . . in providing information, in influencing, coordinating, and affiliating' (Postmes, 2003: 93) are well-attested, and there are features of the collaborative group in particular that make it an interesting subject for study. It is also yields to precise characterisation. Donato, for example, identifies the key characteristics that distinguish collaborative groups from 'loosely configured individuals' (2004: 287):

- a meaningful core activity;
- social relations that develop as a result of jointly constructed goals;
- recognition of individuals as parts of the cooperative activity and acceptance of their contributions in the service of a larger goal;
- coherence in social relations and knowledge 'located and distributed in its members'.

As with Nias et al.'s characterisation introduced in Chapter 1, all of these criteria apply to the three professional groups in this study and influence the nature of their interactional engagement. However, Hargreaves (1992: 226) has suggested that cultures of collaboration tend not to be formally organised and, as the above exchanges show, the achievement of larger goals may effectively be suspended while more local ones are pursued. More useful to this study, therefore, is a definition in terms of group characteristics that will be revealed through the achievement of mundane business, the everyday practices that both respond to and refine common procedures and understandings:

> Collaborative teams ... are characterized by identification with the team, a shared perception of interdependence, low power differentiation, social closeness, collaborative conflict management tactics, and a win–win negotiation process.
>
> (Donnellon, 1996: 207)

The chapters that follow will elaborate the mechanisms that underlie this characterisation and examine some of their institutional implications. One possible interpretive frame that can be applied to this, deriving from social psychology, will be explored in detail in the next section, but a very brief consideration of group conformity in the context of Extract 2.1 will illustrate where analytical disjunctions are likely to arise. Proponents of self-categorisation theory, for example, argue that classical explanations of conformity put too much emphasis on its rational nature. They argue instead that group members process what the group has to say in order to work out what norm the group favours rather than why the norm itself might be sensible. During discussion group members become more aware of their affiliation to the group and therefore make more effort to adhere to prototypic group norms, a stance that is also associated with a wish to distinguish themselves from outgroups (for a brief summary of research, see Baron and Kerr, 2003: 100–02). On the basis of the evidence of the extracts, however, it is hard to account for the group's immediate and concerted response to Helen's action in terms of working out the relevant norm and then adhering to it. Part of the problem is that the 'working out' is jointly, locally and publicly accomplished rather than an individual process reflected in simultaneous private reasonings that become apparent through a process of gradual behavioural adjustment (for a discussion of the relationship between self-categorisation theory and the analysis of talk in interaction, see Dickerson, 2000). From the perspective of an analytical framework that explores local rationalities constructed through ongoing interaction, the group's response is perfectly rational, though not in the way that self-categorisation theory rejects. The nature of the relationship between these different analytical traditions needs to be resolved at the methodological level.

Methodological issues

All three groups in the study have developed ways of interacting that serve to maintain the group as a group through its shared decisions and actions and this, in turn, influences how the group reaches the decisions that help define it. The methodological challenge these relationships represent lies in the way that group processes are analysed.

Traditionally, research on group decision-making has largely depended on the analysis of interaction among members of groups brought together for the purpose of exploring particular phenomena, often under experimental conditions. In his excellent review of work on communication and group decision-making, Frey (1996) includes a summary of his earlier 1988 and 1994 studies which is revealing. The 1988 study revealed that 64 per cent of studies of communication in group decision-making focused on zero-history groups, 60 per cent were in laboratory settings (figures that were even higher in his 1994 review of *Small Group Research* papers), 72 per cent used students and 72 per cent observed groups only once. Where groups were observed more than once, the mean number of observations was only 2.75. The outcomes of such studies are also interesting, producing models or typologies of the variables that influence group processes that, as Poole and Hirokawa (1996: 13) observe, 'seem to have one thing in common: their almost stunning complexity.'

Whatever the value of such work – and complexity of outcomes may represent only a step on the way to a more unified understanding – its neglect of the everyday decision-making of authentic groups over time leaves exposed a potentially serious gap in our understanding. As McGrath and Altermatt have noted (2001), a dependence on experimental groups sacrifices the opportunity to understand the ongoing dynamics of group formation and development, the ways in which groups adapt to and respond to their environments, and how they produce and deliver products. Hence the authors' call for data on group interaction in order to 'find out about these dynamic processes by which such groups do what they do, and modify both themselves and their activities over time' (2001: 526). Interestingly, though, they assume that such an analysis will be based on a coding system.

For longitudinal studies coding systems do offer some advantages, not least the opportunity to compare behaviours over time, but they are necessarily selective in their focus and decontextualised in their representation. The picture they paint may well be suggestive of evolving practices and processes, and this may be the most we can realistically expect, but it cannot hope to capture the intimacies or dynamic construction of the decision-making process itself. For this a different analytical orientation is required, one that will also be indicative of the processes of the group's evolution, just as the visible landscape retains trace evidence of the forces that have shaped it. Attention must be directed to those features of group interaction that have developed distinctive contours over time:

> Most real-life groups are embedded within a history that constitutes and continually is reconstructed by their communication practices and decision-making outcomes. This shared history, constructed socially over time through language, arguments, stories, and symbols, represents a 'deep structure' that influences the 'surface structure' of a group's interactional patterns and decision making.
>
> (Frey, 1996: 19)

Such approaches will treat communication not merely as a medium for decision making but as constitutive of it (Poole and Hirokawa, 1996); the interactional patterns are, in Maynard's terms, the substance, 'the site where people produce elementary forms of social organization' (1987: v). Hence the analysis of particular instances of talk will reveal not only its locally occasioned character but aspects that have been constructed over time and are reflected in current orientations. The group's treatment of Helen's move and its very different response to Jack's introduction of the same topic, for example, are explicable only in the context of accepted procedures and shared understandings built up through innumerable encounters of this kind. There is some evidence that groups develop in-group 'codes' as they interact over time (Cutting, 1999, 2000), but this is only one aspect of a bigger picture of engagement at different levels, and in the chapters that follow the elements in Frey's list will be used as the basis for exploring aspects of this 'deep structure' and their implications.

The call for a study of situated (Potter, 2000: 36) and reproduced (Giddens, 1979: 117) practices and the associated claim that discourse analysis has been underutilised in the study of groups (Stohl and Putnam, 2003: 411) has been taken up by a number of researchers unhappy with cognitivist approaches, generating new ways of understanding group practices. In their description of discursive social psychology (DSP), which reflects these new orientations, Potter and Edwards (2001: 104) identify three key aspects of its discoursal orientation: the fact that it is *situated*, *action-oriented* and *constructed*. Discursive social psychology, they argue, combines an interactional orientation, in which talk is treated as occasioned in the CA sense of the term, with a rhetorical perspective reflecting the different, often competing, positions taken up by those involved. The focus in such work is on how actions are 'pervasively being done' (ibid., 109), achieved *through* discourse and not merely reflected *in* it. As the authors note, this means that cognitive entities, traditionally treated as analytic resources, are approached empirically as participants' ways of talking.

This leads naturally to an interest in aspects of discourse construction, both in terms of its features (stories, metaphors, accounts, etc.) and in terms of the way it constructs versions of the world; as Potter and Edwards put it, 'discourse is both construct*ed* and construct*ive*' (2001: 106). It is this perspective that offers the best response to the recognition that an adequate theory of group decision making must ultimately depend on its ability to reflect the complexities of interaction (Poole, et al. 1996).

Analytically, such an approach must be empirical, developing its arguments directly from the evidence of the data, and it must be detailed if it is to capture tiny aspects of the unfolding interaction that account for participant interpretations. Such detail, and the need to examine sequences of talk rather than isolated examples in order to see how it is built up by the participants, has implications for the presentation of data. As Potter and Edwards note, different studies call for different procedures, sometimes working through single transcripts, sometimes drawing on a large corpus. The data set in this study is large (the Pen transcripts alone run to 400 pages), but in order to understand the nature of the interactional and interpersonal forces at work it is necessary in most chapters to represent these in extended sequences. Occasionally, collections of shorter extracts or examples will be used, but the primary aim of the analysis is to explicate processes in action rather than to present instances to illustrate general characterisations.

In practical terms such studies will contribute to our understanding of group processes and decision-making but their focus on the nature of interaction offers no easy recipe for training interventions. Simple prescriptions for changing the nature of interaction look conspicuously thin when set against the subtleties revealed by close analysis, which is not designed to yield convenient models to guide behaviour. The most the analyst can hope to do, then, is raise awareness of some of the broader interactional forces at work in everyday professional interaction.

Conclusion

All groups develop their peculiar ways of doing things, orienting to signals, structures and routines that define the group and provide the means by which members achieve their goals. This chapter has provided a flavour of the interactional power that the group is able to wield, showing how it responded to a local perturbation by subtly reorienting to normal business. Helen's attempt to adopt – however temporarily – the role of Chair in order to begin a Centre meeting in the absence of the Director is not explicitly rejected by her colleagues; instead, by responding in a way that prevents the topic from developing, they establish favourable conditions for Tony's intervention, which moves the talk away from the agenda item that forms the basis of Helen's bid. The extent to which their talk has been specifically designed to do this becomes fully apparent only when the treatment here is contrasted with the response when the item is reintroduced by Jack.

This is not the way things normally happen in these Tuesday meetings. Jack's lateness is unexpected and Helen's attempt is a spontaneous response to this, not part of a concerted plan that she has had time to prepare and discuss with colleagues, while their response itself, co-ordinated though it might appear, is also spontaneous. Nobody has signalled that they should react in this way to Helen's introduction of the topic and there are no explicit rules about who can begin the meeting or in what circumstances, yet the group are orienting as one to their unstated understanding of how things work, an appreciation built up over thousands of hours of doing business through talk.

Over time, important 'seen but unnoticed' ways of speaking will be developed, traces of which will be evident in the group's daily exchanges. These traces will point to a larger and more complex interaction order to which members will orient and, although the precise contours of each group's interactional landscape will be different, there will be common features. The chapters that follow examine three groups which do not share the same professional ends but are constituted – in a way that might feature in any working environment – as teams of experienced players with shared goals and an overtly collaborative orientation. Paradoxically, while all three groups have proved demonstrably successful in professional terms, they have ultimately failed to achieve a more fundamental goal: their own survival. Three organisations have lost key personnel, the groups' work has been radically changed if not completely abandoned, and the shared expertise gathered over a number of years is now scattered around the globe. Why? Is there any evidence in their talk that might point to a reason for this failure and what lessons might this have for similar groups? If there is any such evidence it is to be found in the '[w]ords, metaphors, idioms, rhetorical devices, descriptions, accounts, stories and so on' that are the staple of discursive social psychology (Potter and Edwards, 2001: 105-6).

This book argues that the interactional inbreeding characteristic of successful collaborative groups produces an unusual amalgam of strengths and weaknesses, reinforcing unity and efficiency at the expense of exposure to internal weaknesses such as groupthink and external threats arising from their relative isolation. The norms and routines that underpin their efficiency, like the strengths that derive from their sense of common purpose and shared beliefs, depend on developing ways of speaking that allow them to circumvent the need for the sort of direct conflict that might leave an individual member isolated and the group thereby divided. In their most sophisticated and corrosive form these interactional rat runs respond to problems or challenges either by resolving them without ever explicitly acknowledging them, or by attributing them to an external agency which is represented as a threat to the group and its activities; when operating more positively they serve as mechanisms for focusing the group's energies, building on shared understanding and producing professional outcomes of the highest quality.

Since the vulnerability of such groups derives directly from their strength, radical excision of 'offending' practices is neither possible nor desirable; instead each group must strike a delicate balance between allowing the individual voice to emerge clearly while at the same time constantly reaffirming the identity of the group as a whole. The chapters that follow examine the groups' interaction in order to provide the sort of understanding that might make such a balance possible.

3
Staying Onside: The Negotiation of Argument

Despite its usually negative connotations, argument in talk is a pervasive feature of our interactional being and one that serves a diversity of ends. 'Social life', as Antaki notes (1994: 160), 'is argumentative'. This chapter will examine the social aspect of argumentative engagement; specifically, it will explore the ways in which a collaborative group appropriates argument and makes of it an affiliative resource. Argumentative involvement could be seen as representing a challenge to the collaborative identity that has been described in the foregoing chapters because it presupposes differing positions, but this chapter will show how interactants develop a centripetal dynamic by drawing on shared understandings and resources in order to engineer collaborative positioning which reorients individual differences within a broader common purpose. I will suggest that this process depends on an orientation where notions of winning or losing an argument have no valency and that, instead, arguments are constructed for the purposes of *bringing onside* those involved.

The idea that argument may have an important social function is not new; writers from very different perspectives have pointed to its pervasiveness (e.g. Jacobs and Jackson, 1982; Billig, 1987; Antaki, 1994) and most have recognised its important function as a regulatory mechanism:

> First, as an essentially dialogic enterprise, argumentation is difficult to discuss without reference to the participants who alternate in taking positions and disputing and supporting these positions. Second, the interactive nature of argument sheds light not only on its structure, but also on its function as a disagreement-regulating mechanism.
> (Kleiner, 1998: 593)

Schiffrin has also shown how argument can serve as a form of talk that is invoked and constructed for the specific purpose of strengthening social bonds (1984, 1990). However, the claim I wish to make here is not that argument is sought out for such specific purposes, rather that where disagreements arise, as they inevitably will in any dynamic professional situation, the interactants exploit the opportunities that these represent in order to reinforce the cohesion of the group. Strategically sophisticated as this may be, I am not suggesting that it is a conscious process or that interactants are aware of the effects of their contributions in this respect. It is best seen as part of a continuing process whereby the group is interactionally co-constructed and where interactional motifs are woven into the fabric of talk.

In order to show how this process works I will offer a detailed analysis of a particular argument in one of the three research settings, using this to develop a model of argument that is very different from the standard representation. This will then form the basis for discussing arguments in the other two settings and a consideration of situations where more oppositional formats occur. Arguments, by their very nature, tend to be relatively lengthy and this presents the analyst with tricky representational challenges: show too little or settle for a summary and interactional complexities are lost, go into too much detail and the range of examples must be so limited that the reader has to take a great deal on trust. Somehow, a balance must be struck. The central example in this chapter will necessarily be relatively long and the others rather selective, but in all cases close analysis will provide the opportunity to examine how interactional positionings serve collaborative ends.

An argument

As arguments go this one is fairly unexceptional. It takes place in one of the group's regular weekly staff meetings, lasts about ten minutes, involves all ten of the people present, and concerns an issue that has not arisen before and will not be revisited. It has been preceded by a 20-minute presentation by the head of the relevant university department (HoD) who has outlined the issues relating to a new European Union (EU) working time directive. This requires workers in certain categories to complete a description of their work every week in order to ensure that they have not exceeded a maximum of 48 hours, and although employees are able to opt out of this system – in which case they can work for more than 48 hours – they are still required to submit the relevant information. Only certain categories of staff are subject to this directive and, because contracts in the Centre differ, some of its members will be exempt while others will be required to submit the required information. Those who are required to do so have expressed their dissatisfaction with what they are being asked to do and when the HoD leaves one of the group challenges her position.

The argument opens conventionally, with Tony presenting a position that he sees as different from the rest of the group. Jack challenges this position and the argument centres on the extent to which the EU requirement represents a threat to the group's way of doing things. Claims are made about its divisive character, the extent to which it undermines a culture of trust and an individual's freedom, and the time required to complete the relevant form. All of these claims are challenged by Tony and provide the main focus of the talk until attention shifts to issues associated with the completion of the form and interpretations of what this involves. Previous claims are revisited but a more explicit theme of external challenge to the group emerges. When Paul suggests that the EU form is badly designed this provides the basis for common ground as the form and the HoD's representation of the situation come in for concerted attack. Both Paul and Jack are able to take advantage of the consensus that develops to restate their earlier claims, now effectively recontextualised. The argument concludes with an agreed course of action: the group will return the forms if they are not happy with what is required of them and, if the forms can be completed in a way that suits their purposes, they will adopt an approach that provides useful information to the group while denying access to outsiders who might use it against them.

Even this very brief summary of the argument brings to light a feature of considerable interest from the point of view of the group's collaborative orientation. The consensus that eventually develops does not arise from any concession, submission or compromise, but from a process through which the oppositional orientation is subtly shifted to an external agent. This both provides a platform for the eventual commitment to a shared course of action and permits the reinvocation of statements that were originally oppositional but are now recast as explanatory by virtue of their recontextualisation. The process of realignment reflected here will be examined in more detail later; for now it is enough merely to note that it is present as a surface feature of the argument.

Negotiating difference

One disadvantage of surface representations of argument such as the one above is that they can provide a misleading impression of the ways in which such arguments are constructed, hiding the interactional subtleties that contribute to their development. Within a collaborative context no argument starts cold: the emergence of opposition must be carefully managed. Because of the way the talk is designed a close analysis of the first 32 lines of the interaction, revealing this process as it works towards the establishment of opposing perspectives, may give the impression that this is not, in fact, an argument. A schematic representation of this short stretch of talk paints a different picture (Figure 3.1).

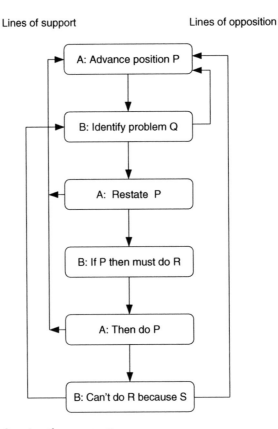

Figure 3.1 Opening the argument

The realisation of the positions that give schematic shape to these opening exchanges establishes more subtle and interactionally potent strategic contrasts than are apparent here. We can see this in the opening turn of Extract 3.1a, a statement that weaves together social and interactional positions:

Extract 3.1a

```
001  Tony:  Am I on my own on this.=Am I the only guy on
002         this page? I don't see Any problem with this
003         at all.=At the end of the working da:y,
004         (0.2)
005  Tony:  you reflect on how many hours you've worked 'n
006         you make a note of it.=
```

Tony prefaces his argumentative position with a statement (with metaphorical repetition) of his orientation vis-à-vis the group, asking whether he is isolated in his interpretation of the situation. By representing himself as someone who appears not to share the common understanding he positions himself outside the group, thus implicitly recognising the identity of the group and holding out the possibility of resolution through the discovery that he has in fact misunderstood the group's position. His formulation of what needs to be done is straightforward, suggesting, but not making explicit, his purported confusion: if things are indeed this simple, it would seem unlikely that others do in fact see them differently. The individual orientation is underlined by the stark contrast between the 'I' of the person making the statement and the 'you' who feature in the description of the actions required.

The sequential structuring of talk that sees one recipient-designed turn forming the basis for the next turn (Extract 3.1b), which builds upon it, is not at first obvious in Jack's response, although much flows from it:

Extract 3.1b

```
007  Jack:  =Yeah the (xxxx) part of the problem is
008         dividing it. I mean there >has been< a lot of
009         strong feeling >but the problem is eh< (.)
010         as it was:
```

The initial 'Yeah' is a direct response to Tony's suggestion, but the apparent agreement is deceptive. As Kotthoff (1993) has pointed out, unmodulated dissent is not usually found in the first dissent turn and the presence of unmarked agreement in the context of an argument, which is what we find here, is likely to signal disagreement later on. The argument context is not fully established, but the grounds for it are laid in Jack's response. At first sight it is difficult to know what to make of this: he refers to a problem of division, presumably referring to 'the working day' rather than the problem itself, then mentions strong feeling, an indexical shift whose force is presumably not lost on colleagues (both this and the division will reappear later in the argument), and finally seems to state that the problem remains as it was. As arguments go this doesn't seem to amount to much.

If the logic of this turn defies analysis its interactional design is more accessible. In fact, it illustrates a feature of argumentative discourse captured by van Eemeren et al. (1993: 172) in their observation that 'a great deal of ordinary argumentation does not take place through straightforward exchange of assertive acts; indeed it often seems to create a *purposeful indefiniteness* of force and meaning' (my italics). Although Jack delineates only vaguely the nature of the problem, the repetition of 'problem is' within so short a turn establishes this as a direct contradiction of Tony's denial that any such problem exists. At the same time the indefiniteness of the contribution allows Jack to withhold direct disagreement, thereby allowing for the possibility of resolution before opposition is explicitly established. In his original formulation of the problem Tony left open the possibility that he might have misunderstood the group's position and although this response suggests that he has, it does not commit either participant to a formal recognition of the fact. Thanks to Jack's purposeful indefiniteness there is so far no overt division within the group.

At a deeper level, though, Jack has established a significant contrast that he will work up as the exchanges develop. Tony ended his turn with a very simple statement of what needs to be done in order to comply with the directive but Jack's complicated turn suggests that the situation is in fact much more complex than this. The reference to 'strong feeling' hints at a significant interactional history and he further suggests that this is not a new problem. Implicitly it is not a straightforward problem either and he must be aware that the others present are the very people who have represented their feelings to him as director of the Centre. This subtle contrast between the syntactic and semantic simplicity of the Tony's articulation and the allusive intangibility of Jack's response is developed as the two positions harden into opposition.

Things begin quietly enough:

Extract 3.1c

```
011         (0.2)
012  Jack:  ⎡⎡(xxxxx)⎤
013  Tony:  ⎣⎣I think⎦ we should be opting in to the scheme
014         not opting out of it.
015         (0.4)
016  Jack:  If we were opting in then we should do it
017         together (.) e:m (.) as it stands it's
018         divisive because ⎡it's dividing (us all)⎤
019  Tony:                   ⎣So:  so  let's-  let's⎦
020         opt in I say
```

Tony presents the issues in terms of a simple binary choice where one option is preferable to the other. This is not directly rejected by Jack but he indicates that a condition attaches to the preferred option – one that is not currently met. Tony ignores this explanatory addition and responds directly to Jack's opening statement ignoring its modality and using 'so' (l.19) to bind his proposed course of action to the recognition that this is indeed possible. The 'so' has no logical force but rhetorically it allows Tony to represent his position as built on that of Jack. Interpreted in this way the decision appears collaboratively established.

Jack's initial response to Tony is brief and clear (l.16) but clarity is sacrificed in his continuation where he again exploits the rhetorical advantage of (near) repetition, disguising this with a shift from adjective to verb and the use of a misleading causal link. More powerful than this, though, is the cumulative force of shared purpose. Jack represents his position as that of the group, a 'we'/'us all' under threat of division, avoiding any suggestion that he is speaking as an individual. This contrasts markedly with Tony's assumption of personal responsibility for each of his statements: his first turn begins with 'I think' and his second ends with 'I say'. In terms of *participation status* (Goffman, 1981) Tony both produces the talk (is the animator) and is responsible for it (the principal), while Jack's representation of his position at least suggests that he speaks for the group. The figures they portray are very different: Tony maintains his statement as the outsider while Jack works towards a representation based on the solidarity of the group.

Running through the literature on argument is the assumption that it is based on opposition, a position most starkly represented by O'Keefe and Benoit:

> One basic characteristic of arguments is opposition; opposition is not, in a sense, any concrete act or series of acts, but rather a type of relationship between interactants. Thus 'having an argument' is characterized by the existence of a particular relationship between participants.
>
> (1982: 162)

As if to underline the fundamental nature of this characteristic, they further insist (1982: 168) that 'opposition is not an incidental property of arguments'. Yet in this argument, as in many others within collaborative contexts, what emerges is not an oppositional relationship but a failure of alignment, a disturbance of the interactional balance upon which shared decisions depend. This suggests that in certain contexts opposition may be merely an incidental property of argument and that what is fundamental is the disturbance of the presumed relationship upon which interaction is built. Once steady state has been re-established normal service can be resumed, although the argument itself will have contributed – however minimally – to a reconfiguration of that normality. Where certain norms are already slabbed within the bedrock of a relationship, as with collaborative groups, these may serve to override any simple oppositional format, influencing the orientation of participants to the unfolding talk. And this in turn will depend in part on how identities are negotiated. Smithson and Diaz (1996) have suggested the need to recognise two levels of identity: the physical speaker and the socially constructed voice. The latter is not susceptible to simple reductionism, as the development of this argument will show.

Jack has already identified himself with the group as a whole and now goes on to cement that relationship by invoking shared understanding:

Extract 3.1d

```
021          (0.2)
022   Jack:  The- we CAn't opt in because e::m (.) we don't
023          have any choice but to opt in an- and with in
024          if- uh-=
025   Paul:  =If (a)=
026   Jack:  =the thing goes through
027          within ┌about a year┐ and=
028   Paul:         └I you're    ┘
```

```
029  Jack:  =a bi:t,
030  Paul:  Yeah=
031  Jack:  =everyone else is >gonna be a lecturer but
032         they're gonna have to< waste their time
033         between now and then filling in forms
034         ┌(which)┐ otherwise they=
035  ????:  └(xxxxx)┘
035  Jack:  =wouldn't have to do.
036         (0.6)
037  Jack:  >Because< it's no good-it's just e:xtra work
038         for people >it's it's< very annoying and it's
039         unnecessary.
```

The opening statement seems to be contradictory, the use of 'because' no more than a nod in the direction of logic. However, the power-lessness represented here and pursued in his description of the effects of an external edict is in marked contrast to Tony's active orientation. In terms of content, these turns seem even more obscure than Jack's earlier ones, although Paul's support in line 30 suggests that not all participants see it in this way. Like the others, he will have understood 'the thing' as a recent decision to consider moving all teaching fellows ('everyone else') to teaching-only lectureships within 'about a year', thus rendering the intervening actions a waste of time. The argument, which ends with a three-part rhetorical flourish (Atkinson 1984), is based not on opposition to Tony's position but on the establishment of a solidarity of understanding which sees the group as threatened by the actions of an outside agency. Tony has proposed that the recommendations of this agency be adopted and, by implicitly attacking the agency rather than the response Tony has advocated, Jack keeps his colleague out of the firing line. The rest of the argument will revolve around bringing Tony onside by a process that will depend on a gradual shift towards a position where he can once again identify with the group.

The development of opposition

There follows a brief side issue introduced by Mark's suggestion (ll.40–52) that the directive would be interesting if it insisted on their working *at least* 48 hours rather than a maximum of 48 hours because people would then fabricate their returns. Jack's response leads the discussion back into its original ambit by echoing earlier references to

feeling and the temporary nature of the arrangement. He then raises
the issue of trust:

Extract 3.1e

```
062  Jack:  We've always worked by trust, we haven't asked
063         people what they do:.
064  Tony:  I don't see any attack on (.) trust in this at all
065         I really don't.
066  Jack:  I do:.
067  Paul:  >Yeah but< what this is takin:g (.) taking away
068         your freedom to: °in° in one particular month,
069         to write a coupla papers that you really wanna
070         write,
071         (0.4)
072  Paul:  and maybe because you (.)'ve not been organised
073         before in: (.) in getting your time
074         (0.2)
075  Paul:  together to write those articles >but you< wanna
076         write those articles ⌈>b'cause it's<⌉ taking away
077  Tony:                       ⌊ Mmm         ⌋
078         (0.3)
079  Paul:  your right
080         (0.6)
081  Paul:  to determine (.) the way you choose to
082         (0.3)
083  Jack:  Yeah.
084  Paul:  to apportion your ti: ⌈me, ⌉
085  Tony:                        ⌊No  ⌋ it's not,
086  Paul:  in your life.=
087  Tony:  =it's taking away the university or your
088         employer's
089         (0.2)
090  Tony:  ri:ght, to insist (.) that you do such a thing.
```

Here, for the first time, there are elements of direct contradiction. Issues
of interpretation are baldly contrasted and the association between individual and statement is direct; the earlier subtle attempts to avoid direct
confrontation have been left behind. Tony's first response (l.64), while
questioning the relevance of Jack's claim, is consistent presentationally
with his opening statement. It recognises that this is a personal position
and allows for at least the possibility that he has failed to see something
the others have taken on board. He underlines the personal nature of the

response at the end of his turn, but Jack's reply to this is brief and stark: they see things differently. At this point (l.66) the participants have moved from attempts to negotiate their differences to argument proper. Jack's use of 'I' here stands in marked contrast to his earlier use of the plural pronoun and is structurally designed to respond to Tony's 'I really don't', setting up an oppositional format, which Tony then exploits in his response to Paul's shift of focus. Although Paul is forming an oppositional alliance with Jack and the placement of his disagreement turn is consistent with the normal pattern for this (Kangasharju, 2002), its form is far from typical. It begins with 'Yeah but' and is developed at length around the idea of 'taking away' one's 'freedom' or 'right'. Tony uses *format tying* (Goodwin and Goodwin, 1987) to counter this by arguing that the directive is designed to take away the university's, not their, right (ll.87–90). This 'verbal shadowing', as Kotthoff notes (1993: 203), 'exploits the resources of one act to construct the next'.

More subtly still, Tony here seems to be hijacking the final element in a three-part rhetorical sequence set up by Paul. Evidence of the power of such sequences in political persuasion has been provided by Atkinson (1984), but their relevance in argument has not so far been sufficiently appreciated. If, as Atkinson suggests, the power of the sequence depends on the climax provided by the third element, the hijacking of this third element not only takes the wind from an opponent's sails but utilises it to propel one's own argument, thus deriving a double benefit from the manoeuvre. In Extract 3.1f we see it used again almost immediately by Mark, once more at Paul's expense:

Extract 3.1f

```
091 Paul: >No but< I'm not a̲l̲l̲o̲w̲e̲d̲
092       to wr⌈ite.  I'm  not  allₙ o̲w̲e̲d̲ to do more=
093 Jack:     ⌊Y̲e̲s̲ (.) that's the-⌋
094 Paul: =than f̲o̲r̲ ⌈ty-eight h̲o̲u̲:̲r̲s̲.     ⌉
095 Mark:          ⌊Well you're allowed⌋
096 Mark: to⌈sa:y    whatₙ on earth you l̲i̲:̲k̲e̲ ⌈about it⌉
097 Tony:   ⌊There you a̲r̲e̲!⌋                [        ]
097 Paul:                                    ⌊unless I⌋
098       opt o̲u̲t̲.=
099 Jack: =Unless you opt out (.) yeah.
100       (0.3)
```

The use of the three-part sequence, then, carries a risk with it but it may be a risk worth taking. Later in the argument it enables a speaker to hold the floor when he has seized it, when Tony's interruption of Helen, which would normally follow the 'no overlap' rule for English conversation (Sacks *et al.*, 1974) and be terminated in the face of the first speaker's determined pursuit of her legitimate claim to the floor, is carried forward successfully on the back of a three-part sequence:

Extract 3.1g

```
180   Tony:   Because some people are working MAny too many
181           hours.
182   Helen:  But then if- ⌈It's that many hours (relying=
183   Tony:              ⌊And it does redound on their =
185   Helen:  =on your memory it is a burden memory=
186   Tony:   work and it does redound on their families
187           and it =
188   Helen:  =because I-          ⌉
189   Tony:   =does redound⌋ in all sorts of
              diff⌈erent⌉ ways.
190   Jack:       ⌊Yes.⌋
```

In the case of the earlier exchange (Extract 3.1f) the strategy is not successful because Mark hijacks Paul's 'allowed to' and is immediately supported by Tony's emphatic 'There you are', designed to underline the force of Mark's assertion. In contrast to the elusiveness of some of the opening exchanges, built up around a delicate choreography of non-engagement, participants at this stage move freely into one another's interactional and semantic space, and contributions are treated as exploitable resources:

l.95   Mark interrupts Paul's turn, creating a third part using the 'allowed to' structure on which Paul's turn was built;

l.97   Paul in turn interrupts Mark's contribution, providing a formulation ('unless I opt out') which would serve to complete his own earlier turn or Mark's, thus qualifying both and reclaiming the position;

l.99   Jack repeats Paul's formulation.

This evidence underlines the fact that arguments in collaborative contexts are every bit as hard fought as arguments in more confrontational circumstances; the participants here sacrifice none of their determination to persuade others of the rightness of their case and are prepared to use whatever interactional resources they can to achieve this end. There is evidence here, for example, that they are engaged in the strategic negotiation of relevant topics, seeking the advantage that derives from establishing topics suited to their own positions (Kotthoff, 1993: 201). An examination of the devices deployed to gain advantage in the discussion reveals at once that the participants are not seeking accommodation with those who hold to different positions. In this particular disagreement Tony finds himself more or less on his own, so a brief list of his strategies will give a useful flavour of moves that interactants can make. The following labels, together with examples, should need no gloss:

Upshot	'so let's opt in I say'	19
Reconstruction	'But this is like saying...'	193
Consequence	'Well in that case...'	234
Challenge interpretation	'I don't see any attack on trust in this'	64
Challenge accuracy	'That's not what was represented to us'	131
Challenge relevance	'That's a different issue'	113
Challenge upshot	'Why are we in trouble?'	216

This is not offered as an exhaustive list or as in any way definitive, although it is fundamentally similar to that provided by Muntigl and Turnbull (1998), differing only in matters of detail. For practical purposes what matters is that Tony's strategies could be described by any of the systems currently available and would be recognised as interactional moves designed to undermine a rival position. In collaborative contexts, as elsewhere, disagreements have to be managed and the resources called upon in order to achieve this will be fundamentally similar. A useful and oft-quoted distinction drawn by O'Keefe captures the significant difference between these local resources and the wider discourse in which they are embedded (O'Keefe, 1977:121):

Argument[1] 'a kind of utterance or a sort of communicative act' which relates to 'he made an argument'
Argument[2] 'a particular kind of interaction' which relates to 'we had an argument'

Tony's moves are cases of Agument[1], examples which might be 'describable apart from the particulars of their occurrence' (O'Keefe, 1982: 18), and we would expect to find examples of such moves in any argument sequence. Using O'Keefe's distinction, if we slip into the assumption that an argument in the sense of Argument^2 is no more than an agglomerate of particular arguments (in the sense of Agument^1), we are likely to ignore more subtle aspects of the design of such talk: examples of Argument^1 will point us inevitably to oppositional features of the interaction, disguising alignment at a deeper level.

Evidence of this emerges in the extent to which participants sought at the beginning of this sequence to present their positions in such a way as to avoid any overt demonstration of opposition. Tony allowed that he may have misunderstood the group's position, while Jack cast his arguments in a form so vague and apparently contradictory that they suggested rather than expressed a difference in orientation. In this opening phase (ll.1–83) there are 14 'positional' turns (by this I mean a turn which might be represented on a diagram representing the propositional development of the argument, as, for example, a statement of a position or a challenge to a previous statement) and of these nearly half (6) begin with 'Yeah', including two that in fact begin oppositional statements. Quite a lot of interactional water has passed under the bridge by the time the first direct contradiction emerges in line 66, but from this point to the end of what might be termed the argument phase (l.246) a more obviously oppositional orientation is established, as evidenced in a reversal of responses: this time, nearly half of the responses (19 out of 42) begin with 'no', 'but' or 'well', and only seven, all supportive of speakers sharing the same position, begin with 'yeah'.

Finding common positions

This oppositional phase is, in fact, surprisingly short. Not long after it has emerged, Jack introduces the agency of others as a potential threat to the group: 'the (.) potential fo::r (0.2) e:rm (.) using those records against us which they could do.' (ll.145–6) prompting Tony to claim that 'we fight it 't that level'. This lays the foundation for the creation of a new oppositional orientation, allowing a shift away from the potentially divisive format that has just emerged. The turning point is Jack's repeated characterisation in ll.247–56 of the potential for exploitation as 'dangerous' and Tony's reiteration of the need to 'fight it'. The rest of the talk, well over half of it, is dedicated to finding common ground for agreement. Here the predominance of explicit agreement is striking: 'yeah' occurs 45 times in 84 positional turns, with 'but' or 'well' appearing only four times in all.

The pervasiveness of explicit agreement is in part merely a reflection of the careful management that will shift the talk away from disaffiliative engagement towards a position where the coherence of the group is implicitly reaffirmed. This is achieved using a number of different discourse moves, many of which are to be found in Extract 3.1h:

Extract 3.1h

```
326   Paul:  There's a cracker on there when she- when she
327          says (.) when she says you know 'n I can't (.)
328          she says it a coupla times about NOT- not wanting
329          to suggest but at one- (.) at one point she
330          quite explicitly >says it would be< a good idea
331          (0.2)
332   Jack:  ⌈⌈To opt out.  ⌉
333   Paul:  ⌊⌊to (xxxxxx)  ⌋
334          (0.4)
335   Jack:  ⌈⌈Yeah.  ⌉
336    Pip:  ⌊⌊hhheh  ⌋ huheh
337  James:  Categorising ⌈what  ⌉ you do,
338   Jack:               ⌊°Yeah° ⌋
339          (0.4)
340  James:  is going to be a very serious problem.
341    Liz:  Yeah.
342  James:  Because=
343    Pip:  =Yeah=
344  James:  = much of it falls under (.) two
345          categories ⌈° (xxxxxx) ° ⌉
346   Tony:             ⌊ at lea:st.  ⌋
347   Jack:  Ye⌈ah⌉
348  James:    ⌊Ye⌋s.=
349   Paul:  =And that's the hassle and that's the waste of
350          time=
351   Tony:  =Y⌈e:s.⌉°(xxxxxxxxxxxxxxxxxxxxxxxxxxxxxxxxx)°
352    Liz:   ⌈Yeah  ⌉
353   Jack:   ⌊Yeah  ⌋
```

This short passage of talk consists of an attack by Paul on the HoD's inconsistency and a comment by James on the problem of categorising their work, with the addition of an evaluative comment by Paul. Theoretically this might have been completed in three turns, but in practice we have 19 turns involving six different speakers. More is at stake here than the mere statement of positions. In fact, the talk manifests at least three features that are typical of this final stage in the argument process, where participants are engaged in finding common ground: externalising the problem, introducing side issues, and collaborative positioning.

Externalising the problem

The talk moved into its final stage with the introduction of a threat from outsiders ('they') and from this point on outsiders and outside conditions are cast in a problematic light. Here we see Paul pointing to the HoD's inconsistency, which he highlights as a 'cracker'. Other references in this closing phase include the vice-chancellor of the university, 'the centre down there' and 'the university', all of which are represented as potential threats or causes of problems. Goodwin and Goodwin have noted how argumentative talk involves interactional positioning:

> Larger corporate entities, incipient teams or sides challenging and answering each other, become visible through the detailed organisation of the talk of the individuals positioning themselves for membership on particular sides.
>
> (Goodwin and Goodwin, 1990: 108)

What makes this talk interesting, though, is the way that such positioning is never allowed to develop. Signs of it are there in the argumentative stage but, by making present 'the other' through the talk, these participants are able to construct this final stage so as to replace the division within the group with a division that allows them to be part of the same team. In Sacks's terms (1972), the members as co-participants in the talk design it so as to establish themselves as Members of the category of 'persons taking the same side in this dispute'.

Side issues

Paul's reference to the HoD's inconsistency is not designed to advance the argument and serves to distract attention away from the locus of disagreement. The same applies to Mark's introduction of the problem of categorising, which is certainly not irrelevant to the topic as a whole and takes up an earlier point of Jack's that the completion of the forms is a 'burden' (l.191). Tony had responded to this by listing the advantages of spending time reflecting on the amount of work one has completed in a week. In the meantime, Paul has shifted the line of argument by suggesting that the form itself is badly designed (l.290), effectively inviting examples of bad design, which are duly forthcoming. These are practical issues to which all can respond supportively. By moving to side issues like this (other examples include categorising thinking time outside the university and lunch with guests counting as work) and invoking external responsibility for problematic elements, the group shifts to less contentious ground and establishes sub-topics around which affiliative talk can be built.

Collaborative positioning

Collaborative positioning can be achieved in various ways, and Jack's completion of Paul's opening turn, following a very brief pause, provides a good example of something that Sacks identifies as characteristic of affiliative talk (1992a: 144–9). As we saw in the last chapter, examples of this are common enough in some forms of collaborative talk, and it is perhaps unsurprising to find evidence of them at this stage in the argument. Here, and in Tony's addition of 'at least' to James's mention of two categories (1.346), we see the participants effectively building together the representation of a position and, in so doing, excluding the possibility of alignment with an opposing stance. Completions of this sort appear elsewhere in the data set, which also contains a striking and very unusual example of this feature (Extract 3.2). It occurs in a staff meeting where changes to a course brochure are being discussed:

Extract 3.2

```
001   Paul:   ...if you could have something hanging out the
002           bhhahck! ohf ihht!
003   Helen:  Dhhh (.) H ⌈EHHEHEHEHE:Hheh!                       ⌉
004   Jack:              ⌊ Well if we're gonna do that,⌋ why
005           don't we just change the O in you:.
006           (0.8)
007   Jack:   to that.
008           (2.0)
009   Helen:  °N::o:::° I think that would be really really=
010   Jack:   =silly.=Okay yeah=
011   James:  HA!⌈hahaha   ⌉⌈hahahah      ⌉ha
012   Kathy:     ⌊Hahahha ⌋⌊              ⌋
013   Jack:               ⌊You c'n say it.⌋
```

Jack's completion in line 10 of a challenge to his own position, especially when so bluntly dismissive, cannot be regarded as in any way typical, but it does illustrate the atmosphere in which such talk takes place, an interactional environment where cognitive as well as syntactic completion is possible (Leuder and Antaki, 1988). In Extract 3.1h this is reinforced not only by the high incidence of explicit agreement but also by the extent to which this overlaps other talk (e.g. ll.338 and 351–3). Overlapping and completion are features of what Tannen has called a 'high involvement style' (1984) and occur throughout the talk, but in this context their contribution serves to underline the extent to which all participants are 'onside'.

As all colleagues move onside, the I/we contrast is replaced by an us/them division, and when the argument concludes it is this that Jack chooses to emphasise (Extract 3.1i). Using another move typical of this final phase, he returns obliquely to his original point but reframes this in terms of agreement in principle with Tony's recommendation, closing the topic with a reiteration of the us/them relationship and the combat metaphor that informs it:

Extract 3.1i

```
481  Jack:   =You know. I- this is I- (.) that's what- (.)
482          bothers me a (.) the division in that. In
483          PRINCiple I agree >with the idea of< forty-
484          eight hou:rs: an' I think th't (.) maybe it
485          would be a good idea if we all informally
486          agreed that we would DO THAT. That we w'ld- (.)
487          as you say, quickly reflect at >the end of the
488          day< how many hours we've ⌈spent⌉ that day,=
489  Tony:                             ⌊Yeah ⌋
490  Jack:   =and get s:ome sort of picture >'f how many
491          hours< we're doin', but what I object to (.)
492          is giving them that in the form that they want
493          it.= An' giving them any ammunition to use
494          against us.
```

The analysis presented above suggests that we need to rethink our view of argument, at least as far as collaborative groups are concerned. To see it as 'a kind of interaction in which two or more people maintain what they construe to be incompatible positions' (Willard, 1989: 42) where the point 'is to convince the other side to change its mind' (Blair, 1987: 191) is to direct attention away from much of the interactional work that is being done. It would also be a mistake to suggest that this sort of talk can be categorised as 'sociable argument' (Schiffrin, 1984), designed to contribute to the solidarity of relationships (Schiffrin, 1984, 1990; Georgakopoulou, 2001). The centrifugal force of the disagreements emerging here represents a threat to the coherence of the group and resolution will depend on generating sufficient centripetal interactional energy to draw it back to its core. Hence the argument is co-operatively designed in order to lessen the force and extent of the emergent disagreement and maximise opportunities for finding common ground. What matters here is not so much winning the argument but bringing colleagues onside.

A trajectory for collaborative argument

Argument is one of the foundations of professional collaboration. Keeping this apparent paradox in mind directs attention to the dynamic nature of collaborative identity and the extent to which it depends on reconciling the individual energies of its members. It is hardly surprising, then, that the above analysis suggests a trajectory for collaborative argument that is very different from conventional descriptions. This section will sketch the path that such a trajectory might take, comparing it with a standard description and briefly indicating some of the interactional issues that are involved. The next section will then explore aspects of collaborative argument on the other two sites in terms of this proposed trajectory.

The model of argument used for the purposes of comparison (Figure 3.2) is that suggested by van Eemeren et al. (1993: 31), based on the 'functional stages' of an argument. The two descriptions share a common argumentation section although, as we have seen, in practice this can be relatively short in the case of collaborative argument. And where the ordinary argument depends for its point of departure on confrontation based on the expression of a standpoint, collaborative talk seeks to minimise this, instead involving the participants in a delicately elaborated negotiation of difference. Similarly, where the concluding section of conventional argument is expressed in terms of the original standpoint, in the case of collaborative argument it is more important to establish a common position first, so that what emerges is a formulation of the original standpoint that all participants can adopt as if from a shared perspective.

The first phase seems particularly important in terms of the evolution of the argument – or its avoidance. Jacobs and Jackson (1981: 123), drawing on O'Keefe's distinction, characterise this as an Argument[1] in the absence of Argument[2], where the argument may be 'projected or implied, but never openly occurs' and it is perhaps worth noting that a move of this sort can occur only in the interactional environment of a 'negotiating difference' phase. Another way of characterising the

Ordinary argument	Collaborative argument
Stage 1: Confrontation • expressing standpoint • accepting or not accepting standpoint	
	Stage 1: Negotiating difference • purposeful indefiniteness • I/we orientation • metaphorical formulation • empty causal connections
Stage 2: Opening • challenging to defend standpoint • accepting challenge to defend standpoint • deciding to start discussion agreeing on discussion rules	
Stage 3/2: Argumentation • advancing argumentation • accepting or not accepting argumentation • requesting further argumentation • advancing further argumentation	
Stage 4: Concluding • establishing the result • accepting or withholding acceptance of standpoint • upholding or retracting standpoint	*Stage 3: Establishing a common position* • collaborative positioning • externalising the problem • focusing on side issues • us/them orientation • pursuing a metaphor

Figure 3.2   Comparison of ordinary argument and collaborative argument

moves suggested in the above trajectory is to view the first phase as one in which there is a search for accommodation and the second phase as arising only when this has proved unsuccessful. Once the argumentation stage has been reached it is necessary to work towards alignments that will allow, if not a return to accommodation, at least the reconstruction of a shared perspective. The success of this depends in part on the nature of argument as an unfolding and interactively constructed phenomenon. As Smithson and Diaz note in their discussion of the 'collective voice' in this context:

> The way participants argue is dependent in part on the current state of the argument, and is not all predetermined by the opinions participants originally brought to the argument.
>
> (1996: 254)

It is important to stress, however, that collaborative argument in professional contexts does not seem explicitly designed simply to reinforce mutually held beliefs in the way that is typical of pseudo-argument (Kleiner, 1998). There are, in fact, a number of important differences between the two forms. Collaborative argument, for example, does involve genuine disagreement, which emerges in the argumentation stage and at least serves to ratify the importance of the topic as something worthy of genuine debate. Similarly, in externalising the problem, participants in collaborative argument do not import an 'absent antagonist' into the talk in the way that is typical of pseudo-argument. In the case of the latter this is not something that has to be worked for and which appears in the final stage but an imaginary opponent set up to be disposed of by arguments framed in terms of the shared ideology of the group. Nor is there any evidence in the collaborative argument in this data set of participants making an effort to present themselves to each other in a positive light.

Collaborative argument, then, shares some features with conventional argument but is designed differently, not as a form of pseudo-argument but as a form of engagement that moves from accommodation to alignment via genuine disagreement. The stages identified manifest certain interactional features recognisable within the talk itself and in the design of turns. In the first phase contributions are built on prior turns in ways that allow interactants to develop different but not opposing positions. The following exchange (Extract 3.1k), where Jack uses an element in Tony's turn to develop his own claim, is fairly typical:

Extract 3.1k

```
011  Tony:  I think we should be opting in to the scheme
012         not opting out of it.
013         (0.4)
014  Jack:  If we were opting in then we should do it
015         together (.) e:m (.) as it stands it's
016         divisive
```

In this stage, as in this example, most transfer takes place at transition relevance places, turns are not completed by others and overlap is not particularly common. This changes in the second stage, where competitive overlap and completion emerge (as we have seen in the examples of three-part structures) and where elements in one turn are appropriated for oppositional reformulation in subsequent turns. In Extract 3.1l, for example, we see Paul's turn interrupted by a supportive utterance (l.82), interrupted again with an expression of disagreement before it is complete (l.85), and finally reformulated in a way that contradicts the original claim:

Extract 3.1l

```
075              write those articles
076              ┌>b'cause it's<┐ taking away
077  Tony:    [  └ Mmm         ┘
078              (0.3)
079  Paul:   your right
080              (0.6)
081  Paul:   to determine (.) the way you choose to
082              (0.3)
083  Jack:   Yeah.
084  Paul:   to apportion your ti: ┌me, ┐
085  Tony:                         └No  ┘ it's not,
086  Paul:   in your life.=
087  Tony:   =it's taking away the university or your
088              employer's
089              (0.2)
090  Tony:   ri:ght, to insist (.) that you do such a thing.
```

Finally, in the last stage (Extract 3.1m), we see co-operative overlap (l.338, ll.347–8) and interruption (l.343), and co-operative completion both before (l.346) and after (l.349) the end of the turn:

Extract 3.1m

```
337  James:  Categorising ┌what   ┐ you do,
338   Jack:               └°Yeah° ┘
339              (0.4)
340  James:  is going to be a very serious problem.
341   Liz:   Yeah.
342  James:  Because=
343   Pip:   =Yeah=
```

```
344  James:   =much of it falls under (.) two
345           categories ┌° (xxxxxx) °┐
346  Tony:               └at lea:st. ┘
347  Jack:    Ye ┌ah┐
348  James:      └Ye┘s.=
349  Paul:    =And that's the hassle and that's the waste
350           of time=
```

These interactional moves realising individual but not contradictory positioning in the first phase, destructive appropriation in the second and
co-constructed shared positioning in the concluding phase (the last two
conforming closely to the differences between formal and substantial
co-operativeness proposed by Gruber, 1998) reflect the overall trajectory
of the argument and its direction, and it is now time to examine arguments from the other research sites in terms of this proposed model.
In the case of the Pen, the trajectory is not difficult to discern, and an
example will be offered as a further illustration of its features, but DOTS
shows certain distinctive differences which illuminate other aspects of
arguments within these groups, revealing them in interesting relief.

Illustrating the trajectory

In the argument to be examined here, 'bringing onside' depends in large
part on the extent to which one of the participants is able to bring
others round to his way of conceptualising the problem. The subject
which gives rise to the disagreement is typical of discussions in these
meetings, focusing as it does on matters of everyday procedure. It is
raised by Harry, who is chairing the meeting: 'Pooling (.) tea::ching
materials. Other than classes.' Perhaps because it is so typical and is
introduced as a formal agenda item, the *negotiating difference* phase is
very brief and there is no evidence of any effort to sideline the debate
before it arises. Even so, there are at least three elements typical of the
first phase, as Extract 3.3 reveals:

Extract 3.3a

```
015  Annette:  And also Ed suggested something yesterday
016            whe:n er >I'm not sure who else was here
017            maybe it was just< (Jenny) and I, e:m
018            about (.) a system of e::r
019            (0.5)
```

```
020  Annette:  putting photocopie:s (.) in a: big box or
021            something.°You°
022       Ed:  There it is. Behind you. It's materialised
023            into the blue box.
024            (0.5)
025       Ed:  >I mean< the one drawback I think (.) i::s
026            that most of our (.) work is >kind of<
027            textbook based. And there's plenty of staff
028            copies around. E:m I don't think that I've
029            been making-there's very much materials
030            type been going on really. °So I - just°
031     Paul:  I- I think what we- a- a good way of looking
032            at it would be: (.) >would be< maybe
033            different strategies with materials.
034            'What I do with this piece of material.'
035            (0.5)
036     Paul:  E:m
037            (0.5)
038     Paul:  or 'What I do with this kind of idea, I
039            think, you can imagine what that box is
040            going to look like after (.) two months if
041            we actually start putting in it. You know
042    Keith:  HEHEha!
043     Paul:  It's just going to be a heap of things. SO I- I
044            thought it- it might be an idea just to-
045            (0.5)
046     Paul:  just to look at different ways of
047            exploiting a piece of material which
048            (1.0)
049     Paul:  which we may know about, and sort of pooling
050            ideas rather than pooling pieces of paper.
051  Annette:  Uhuh
052    Harry:  Although there are probably some (bits of
053            paper that)
054     Paul:  YEAH well yes okay.=
055  Annette:  =Yes.
056     Paul:  But >I mean< bringing that
057            element ⌈in as well.
058  Annette:         ⌊Because both Jenny and I were
059            saying that ...
```

Annette begins the discussion by referring to an idea that Ed has put forward and he accepts responsibility for this but points to a drawback arising from their – or, rather, his – working practices. His turn shows evidence of the *purposeful indefiniteness* noted in the IEC argument and, although 'E:m I don't think that I've been making-there's very much materials type been going on really' is not as incoherent as Jack's contribution, it is difficult to see exactly how it relates to what he has just said. Notice, too, how he yields up the floor, like Annette (l.21), with a noticeably quiet utterance following a transition relevance place in his own talk (l.30), allowing his turn to end on a distinctly non-confrontational note. The lack of any action to complete this statement also effectively reduces the initial 'so' to an *empty causal connection*.

When Paul takes the floor it is in order to disagree with the box idea, although his disagreement emerges clearly only at the end of his extended turn and is prefaced with hesitation markers, hedging and evidence in support of his eventual position. On one level the assertion about the state of the box, which is presented in the form of an appeal for his colleagues to visualise the results themselves, is the justification for his claim that it is better to pool ideas than materials, but a more subtle positioning emerges from his formulation, one which appeals to a shared commitment to an active rather than passive orientation in their work. This contrast emerges at various points in the talk of this group and lies at the heart of Paul's *metaphorical formulation*, which begins emphatically and ends with a subtle transformation. The emphasis on the active 'do' in 'What I <u>do</u> with this piece of material' is underlined by the immediate repetition of the phrase 'what I do' and contrasted with the passive 'heap' that will result from putting materials into a box. At the end of his turn he takes Harry's opening formulation and divides it into two parts that reflect the division he has now established: the alternative to his suggested active and purposeful exploitation of a 'piece of material' is the relatively passive pooling of 'pieces of paper'. In teaching terms a piece of material is something that is deliberately selected, while pieces of paper are mere objects, candidates for the litter bin. The success of his position will eventually be confirmed as others signal their move onside by adopting this representation.

This will develop through all three phases of the argument, but it is interesting to note again how briefly the argumentation stage features in the overall discussion. It can be pinned down to a mere 97 lines (Figure 3.3), followed by over 1000 in which a common position is established, and the creation of this position will draw on the same

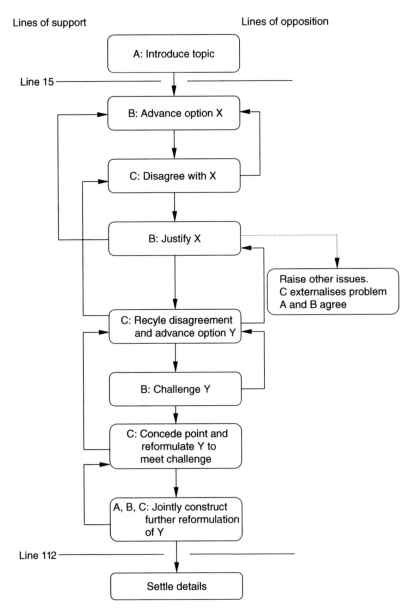

Figure 3.3 A Pen argument

resources as those exploited by the IEC group. The most pervasive aspect is the extent to which Paul's active/passive motif is taken up by the others and woven into the fabric of the discussion. In Extract 3.3b we see an early representation of it as 'a sump', which returns, slightly reconfigured, much later in the talk from the mouth of another participant in Extract 3.3c:

Extract 3.3b

```
086        Paul:  Yeah. And I think that is a sort of a danger
087               >you just< end up with a sump=
088      Harry:  =Yeah=
089        Paul:  =of material,
090    Annette:  Yeah=
```

Extract 3.3c

```
823  Annette:  It should be emptied fortnightly.
824            (1.0)
825  Annette:  Because those things should be: (.)
826            fi:led or discussed (.) or both. A- at the
827            meeting.
828    Harry:  (xxxxxxxxxxxxxxxxx) discover all sorts
829            of things in the bottom of it.
830  Annette:  Yes.
```

There is a sense here in which, to use Boden's terms (1995: 89), Paul has succeeded in drawing Annette and Harry into his logic. Also threaded through the talk is an externalising of the problem, which is introduced early on in the argument (Extract 3.3d) and taken up when further opportunities present themselves (Extract 3.3e):

Extract 3.3d

```
061        Paul:  Except that system up- up the:re is:
062               (0.5)
063        Paul:  who is responsible for keeping it up to date.
064      Harry:  That's the problem=
065    Annette:  =Yes=
066        Paul:  =That's the problem.=
067    Annette:  =Because we haven't got an academic
068               manager.
```

Extract 3.3e

```
194  Paul:  And maybe we're >sort of< (.) getting into:
195         half cock systems rather than
196         (0.5)
197  Paul:  BUT we nee:d (.) whatever happens we need
198         people to: to >sort of< er look after (.) the
199         material
```

Again, this is an issue that emerges periodically in staffroom talk. The post of academic manager was effectively a half-timetable allowance assigned on a rotating basis in order to ensure that one of the group could dedicate time to tasks that would support the teaching of colleagues. It was abolished without consultation in the year before these recordings were made and the Pen staff saw the move as undermining their professional activities. In the argument this form of externalisation is not accompanied by an explicit us/them orientation but its invocation carries with it associations with other negative impositions from 'them' at the other school.

In addition to these dominant themes there are many occasions where attention shifts to side issues – usually focusing on procedure or standard practice – that allow expressions of agreement and support. There are also examples of overlap and completion, as in Extract 3.3f, where Paul and Harry begin at the same time and Paul drops out only to return with a latched completion (1.507) of Harry's claim, prompting Ed's agreement:

Extract 3.3f

```
504   Paul:  [[Well we're go- we ch- we choose]
505   Harry: [[We- we don't go around looking ]for a piece
506          of material to fill a gap=
507   Paul:  =Which is why these things are never used.
508   Ed:    Yeah.
```

This argument, then, follows a trajectory which matches the IEC exchanges and features most of the elements detectable there. Of course, not all arguments take up as much space as this and some elements will occasionally be missing or emerge as particularly dominant. On the whole, though, what emerges from a study of argumentative talk in these two settings is a remarkably similar picture and one that suggests transfer from one setting to another would be unproblematic. In fact, there is an even more striking similarity but to appreciate this we need to look at the DOTS setting where some different characteristics emerge. In order to prepare for this it will be worth standing back briefly for an overview of argument in all three settings.

Different lines

The difference in trajectory between argument as traditionally viewed and its realisation in these two collaborative settings is clear enough, but the situation in DOTS is more problematic. Team meetings there are less regular than in the other settings, but they take pretty much the same shape: the Head of Unit chairs the meeting and leads discussion on matters of current concern, ideas and information are exchanged, actions and procedures are agreed. However, the nature of disagreement here is strikingly different and there is an absence of the sort of extended argument that characterises IEC and the Pen. An example of this difference is to be found in a meeting where a sensitive staffing issue has arisen.

John is often late for such meetings and this one is no exception, so colleagues fill the time with light-hearted banter, which continues briefly into the broad purpose of the meeting: Personal Appraisal and Development (PAD) arrangements. He explains that, having just returned from a training meeting on this subject during which he took 'copious notes', it now falls to him to 'cascade this'. He will arrange a future meeting on the topic but wants now to 'quickly touch on' one or two things. He begins with the subject of merit promotions, a new concept and one which Julie, his deputy, asks him to clarify. His attempt to do this by explaining that it replaces an earlier designation system produces immediate confusion about who is involved in this and a dissatisfied response from Penny, who claims to be still on probation. Just as things are looking sticky, the fire alarm rings and the meeting is suspended. Temporarily saved by the bell, John resumes the meeting by saying that he doesn't 'think there's any point in going any more through the details at the moment', but Julie is not happy to leave the issue of merit promotion unclear. What follows can be summed up fairly briefly: following a summary of the old system John summarises the reasons for scrapping this approach and explains the new one to Julie's satisfaction; Joe then raises the problem of their not having been informed of any changes and having been left with the impression that they are on probation, something that surprises John and leads to a series of exchanges where the interactants attempt to pin down the precise state of affairs while expressing their dissatisfaction with the treatment they have received.

It is a fairly straightforward matter to point to meetings at the other
two sites where the head explains new systems and where colleagues
express their dissatisfaction with the treatment they have received from
the organisation that employs them, but nowhere in the data are
exchanges such as those in Extracts 3.4a and 3.4b to be found:

Extract 3.4a

```
016   John:                                    ... Sue and
017         Penny aren't. On merit promotion]  °Mm°
018   ????: (xx ⌜xxxxx)
019   Joe:       └No, Penny is.
020   John: No°she's not.°
021   Julie: °Yes she is.°
022   Penny: I'm lost.
```

Extract 3.4b

```
250   Mary : Yeah but you should have kno:wn or- or we
251          should have known all through this period
252          that we were on proba⌜tion.
253   Joe:                         └Mm
254   Penny: Yeah.
255   John:  W- ye- no- you we:ren't.
256   Joe:   We were.
257   John:  No, you weren't.
258   Joe:   We:ll,
259   ????:  °Thh ⌜hh!°
260   Joe:        └they were the words that were used.
261   John:  You were o- you-
262   Sue:   No:: (xxx ⌜xxxxx for that.) ⌝
263   John:         └Well I don't think┘ you were
264          be ⌜cause⌝
265   Julie:   └Well ┘it needs confirming then
266          doesn't it.
267   Penny: Mmm=
268   John:  =I don't think you we:re.=
```

In contrast to the approach in the other settings, where disagreement is
something that evolves, where participants seem at pains to represent
themselves as being on the same side, and where disagreement must
be hedged and delicately negotiated, we have here direct and undiluted

contradiction extended over a number of turns. As a number of conversation analysts have noted (e.g. Pomerantz, 1984a; Sacks, 1987), operating within ordinary talk is a preference for agreement and certain turns (e.g. assessments, invitations) are actually designed to elicit a particular response. Where such a response is not forthcoming, the dispreferred response will be marked in some way, with fillers, hesitations, hedges, etc. (for examples, see Levinson, 1983: 332–9). Unmarked disagreement is relatively rare except in children (for an example of extended unmarked contradiction see Maynard, 1985: 3) and there are no examples of extended disagreements in the data from the other sites. In fact, one of the most striking moments in IEC meetings occurs when one participant utters a loud 'No!' in response to a colleague's statement, a face-threatening act that prompts a silence so sudden and complete that it seems to extend into the elaborate repair which follows.

This is not to say that there is no evidence of the unmarked 'no' in response to a speaker's statement, but the environments in which this is found tend to be where the group is working together to settle issues of fact, not as part of the development of an argument. The only exceptions – rare as they are – tend to occur when an outsider becomes involved in a disagreement and, where such a person resorts to this tactic, the results can be overtly confrontational, as the exchange in Extract 3.5 demonstrates:

Extract 3.5

```
001          (5.0)
002  Jenny:  And Kate will submit a list.
003   Kate:  No she wo:n't. She's no idea what she's
004          written.
005          (0.5)
006  Jenny:  That's (.) extremely unhelpful.
007   Kate:  I'm not (.) being
008          ┌very helpful.┐ >I mean< it's=
009   Paul:  └ hhh hhh hhh ┘
010   Kate:  =much (.) better that you people should be
011          doing it °than me.° Anyway these are (.) all
012          things that I've had to do (.) on the back of
013          an envelope prior to the five minutes' notice
014          because (.) nobody else would.
015  Jenny:  Mmm
```

This extract repays brief examination because of the way Kate, overall principal of the two schools in the organisation and regarded by the Pen staff as firmly in the camp of the other school, is prepared to membership herself as an outsider. The significance of this is apparent in Paul's fairly typical view of the other school: 'We don't really have a relationship with Inkham. It's one built on mutual distrust.' When Jenny states that Kate will submit a list of her relevant publications and presentations for a coming review, Kate immediately denies this, prompting criticism from Jenny (l.6). Kate's next turn builds on Jenny's accusation of unhelpfulness and provides a reason why she should not be helpful in this respect. She begins by claiming that the Pen staff should do it and not her, and then goes on to suggest that she has been forced at short notice to jot down ideas because 'nobody else would'. In this single turn she manages to separate herself comprehensively from the Pen group, first by categorising them as 'you people', a group designation that explicitly excludes her, and then by identifying herself as someone who has been obliged to undertake work that nobody else (i.e. nobody in this group) would do. Kate, then, is an outsider by designation and through her actions.

We have seen how a collaborative group can exploit arguments as an affiliative resource, working towards bringing members onside so that the group can present a united front in the face of potential threats from outside but, where one of the participants is in fact an outsider, this changes the configuration significantly. No longer, for example, can the process of bringing onside serve to reinforce the coherence of the group, at least not unless an appropriate accommodation for the outsider can be negotiated. Membership, of course, is a negotiable asset, and it may in any case serve the interests of a participant to invoke categories other than the group, but where this shifts membership to a category perceived by the group as oppositionally oriented there will be implications for the development of subsequent talk.

A close examination of the DOTS meeting in which the exchanges in Extracts 3.4a and b occur reveals that John's membership of the group is by no means a straightforward issue and that his position seems to shift as the talk develops. He actually opens the discussion (Extract 3.4c) by establishing himself as *the odd one out*, without suggesting, as Tony does at the opening of the IEC argument, that this is in any way problematic and in need of resolution:

Extract 3.4c

```
001  John:  Everyone HEre is on merit promotion
002         arrangements °except me°
```

Despite this, his next move (Extract 3.4d) sees an implicit shift to *member of the group* by establishing a we/they orientation through reference to 'they' as the decision-makers in organisation:

Extract 3.4d

```
083  John:  WELL-, all that's happened basically is that
084         they've scrapped
085         (0.2)
086  John:  the designation system
```

Almost immediately, though, there is a further shift away from the rest of the group as his description of the circumstances behind the company's position reveals him as someone with inside knowledge by virtue of his *having senior status* within the organisation (Extract 3.4e):

Extract 3.4e

```
090  John:  but- °·hhh° e::m
091         (0.2)
092  John:  there's been a LOT of PREssure to build in a
093         (0.4)
094  John:  °a° more formal sort of pyramidal structure at
095         (0.2)
096  John:  the research science level
```

This reference to the structure recalls his introductory comments where he informed the group of his responsibility to 'cascade' the information down. These references serve as a reminder both of the hierarchical nature of the institution and, since he is cascading to *them*, of his own position above the rest of group. However, later allusions, such as the one in Extract 3.4f, to his recent relatively long absence because of illness also remind participants that he is an *individual at the mercy of personal circumstances*:

Extract 3.4f

```
178  John:  Well I hadn't (.) well I mean I've (.) lost
179         track >of this bit< but MY:
180         (1.0)
181  John:  u-undersTANding was that-
182         (0.8)
```

```
183  John:  an- <and now certainly I was< asked this:
184         (0.2)
185  John:  in the autumn °I think it must have been the
186         autumn°
```

Finally, towards the end of the discussion, his repeated assurances that he will take up the matter of probationary status, underline his position as *Head of the Unit* and mediator between the group and the organisation, albeit in sympathy with the group (Extract 3.4g):

Extract 3.4g

```
210  John:  I'll take that up then, but that's:
211         (1.0)
212  John:  that's nonsense.
```

These different and sometimes competing perspectives appear throughout his extended turn that sums up the discussion. The opening dozen lines (Extract 3.4h) are fairly representative:

Extract 3.4h

```
284  John:  Y'see y'had a- y- we had a choi:ce,
285         (0.4)
286  John:  Pe- People that worked
287         (0.6)
288  John:  who-=wh- whose manager thought that- that-
289         (0.2)
290  John:  they didn't still: (.) they didn't deserve
291         the:
292         (0.6)
293  John:  current designation, could then go
294         through the=the redesignation procedure
295         o:::r=alternatively, they were given e- (.)
296         up to a yea:r (.) I think
297         (1.5)
298  John:  to review the situation. So we didn't- it-
299         (.) they didn't have to say YEs:, he's: (.) he
300         can keep that designation,
301         (0.2)
302  John:  a-and
```

```
303            (0.2)
304   John:    that can be >converted to< promotion no:w,
305            (0.8)
306   John:    >and I was-< I had a choice of three things...
```

The first line here sees a shift from a 'you' to a 'we' orientation, but this is followed, after a short pause, with a reference to 'people' as a category separate from the management of which he is a member (l.286). This summary concludes with a possible shift from 'they' as 'people' (l.290) to 'they' as people + manager (l.295) and the situation is rendered even more ambiguous by a subsequent repair of 'we' to 'they' (l.298–9), where 'they' might refer to people + manager or the company, though lines 299–306 suggest the latter. He then shifts to an unambiguous 'I' orientation.

In its tortuous imprecision of reference this passage recalls some of the opening positioning in the IEC talk but here it appears as a summing up that underlines the shifting alignments that inhabit John's talk. He is not unequivocally a member of this group so the argument cannot develop here in the way it does in the other centres, and the flat contradictions identified in Extracts 3.4a and b are not as out of place here as they might have been in the IEC argument. There may, of course, be other more particular reasons for the extended sequences of contradiction, perhaps arising from the scientific background of these participants, where matters of detail carry particular weight, but the significance of the sequences lies in their presence in the talk rather than in whatever motivations might lie behind them.

This said, not all aspects of this talk are different from those of the other two centres and particularly striking are the ways in which the organisation is invoked as an external and usually less than competent participator. The conclusion of the discussion (Extract 3.4i) stands as a good example of this and as a valuable reminder that some aspects of argument are indeed common to all three collaborative groups:

Extract 3.4i

```
344   Julie:   (So they're catching
345            you up ⌐now
346                  ⌊((General laughter for 4.0)
347            (3.0)
348   John:    Okay I'll chase that up. °(I shall-)°
349            (0.6)
```

```
350   John:   I suspect this is one of those things
351           that's got
352           (0.6)
353   John:   low down in the list of priorities.
354           (1.2)
355   Penny:  It's just got left in Bellford.
356   ????:   HHheHHht!h
357   Mary:   °They've got it.°
358           (2.0)
359   Mary:   They were on the wrong headed notepaper.
```

Conclusion

This chapter has proposed a trajectory for argument in collaborative settings that is different from conventional models of argument and has suggested that this can be accounted for by the collaborative orientation of members. At this point it is perhaps worth emphasising that what has been suggested is perfectly consistent with what is common to all argument:

> The design of argumentation is assumed to be fundamentally instrumental in nature, that is, aimed towards the practical end of discovering the basis for collective action, agreement or understanding.
>
> (van Eemeren et al., 1993: 142)

It may be that research so far has naturally focused on one particular aspect of argument, defined by van Eemeren et al. as 'argumentation', and that there are many different ways in which this might be embedded in disagreement talk. As we have seen, there are differences between situations where everyone is part of the same collaborative group and those where a member or members may have different loyalties, but the distinction between internal and external, however invoked, seems to be fundamental. In their work on meetings Bagiela-Chiappani and Harris (1997) suggested that in hierarchical models faultlines are to be found between the chair and the group, while in co-operative models they tend to emerge between different groups in the meeting, a finding in line with that of Kangasharju (1996). To some extent the former is true of the DOTS meeting but the latter claim is not borne out where the whole group sees itself as a single unit distinct from potentially threatening outside groups. Where this is the case, the object of the talk is to 'bring onside' those involved and as a result the internal coherence and external validity of the argument itself may be subservient to a higher imperative: social coherence. In his discussion of conflict in decision-making, Putnam (1986) identified three sorts of disagreement: affective, involving interpersonal disagreements; procedural, centring on how collaborators should work together; and substantive, where the focus is on alternatives. It is perhaps not surprising that only the last of these features in the data, the only form, according to Putnam, that is likely to enhance collaborative decision-making. In fact, one might characterise the interactional moves that shape collaborative argument as representing a *logic of affect*, structured rationally in so far as they build towards a conclusion subscribed to by all, but not constructed according to rules of ordinary logic. There are other strategies that can be used to avoid or deflect the force generated by direct opposition (see, for example, Tracey and Ashcroft, 2001; Makan and Marty, 2001, Chapter 2; Craig and Tracey, 2003), but the analysis of these groups indicates that such strategies are subservient to this overarching orientation.

The importance of shared decision-making through collaborative argument has been underlined by the findings of a recent study of the use of oral communication in the workplace (Crosling and Ward, 2002). Although direct persuasion was not seen as a primary need by recent graduates of business schools, the study did point to the need to widen the range of the oral skills taught to business students. The authors also emphasise the 'censured environment' (Thomas, 1995) of the workplace, where sensitivity to local norms and conventions is essential for successful participation. This is what makes the study of different forms of argument so important. There will always be a place for research that enables us better to understand and deploy formal features of argument structure (e.g. Sillince, 1995), but in a professional environment a grasp of sound argumentation is only part of the picture. Knowing how to argue, like knowing how to tell a joke, is a powerful but very dangerous interactional weapon that is likely to blow up in the user's face unless they know just how to deploy it, and in an examination of humour and stories in the workplace we will reveal just how delicate that process of deployment may need to be.

4

The Joke's on Them: Varieties of Humour in Collaborative Talk

> Harry: The amount of deliberate misunderstanding that
> goes on in this staffroom=
> Annette: =Heheh=
> Keith: =HEHEH HEHEH!
> Harry: Heheh
> Paul: What do you mean by that.

Embedded in, arising from and flowing through many of the routines and rituals of professional life, the stream of humour is one of its most distinctive features, providing not only a source of enrichment and nourishment but also serving as a defining characteristic of the territory as represented by its inhabitants. While the last chapter revealed how collaborative professional groups minimise the impact of centrifugal forces generated through argument, this one will examine how the centripetal dynamic of humour is exploited in different ways by such groups to a variety of affiliative ends.

The chapter begins by establishing a brief working definition of humour that will serve as a basis for the analysis that follows. It then considers the view that humour serves as a collaborative resource before moving on to highlight interesting features of its occurrence in the three settings being studied. What will emerge from this is a broadly similar picture of its function within formal meetings but significant differences in encounters during 'free time'. Once these have been highlighted attention will shift to the Pen staffroom, where a detailed examination of humour reveals a more subtle interplay of professional and interpersonal elements than has so far been appreciated. As Holmes has noted (2000b: 161), there is to date very little research on humour in the workplace drawing on recordings of authentic exchanges and, if nothing else, the evidence of what follows indicates that this neglect represents a wasted opportunity to better understand professional worlds.

Humour: a working definition

Not everyone finds humour amusing. The journalist Malcolm Muggeridge, for example, once claimed that good taste and humour are a contradiction in terms, 'like a chaste whore' (*Time*, 14.9.53). Matters of taste are of less immediate concern to analysts of humour, who have first to wrestle with the knotty problem of pinning down the subject of their study. The criterion that most immediately suggests itself as a candidate is laughter, and some analysts have used this. Fillmore, for example, extends the concept of humour to embrace friendliness or good-naturedness and treats laughter 'either as an acknowledgement that something humorous has just been said, or as a prelude to the laugher's decision to interpret what has just happened as an occasion for linguistic fun' (1994: 272). Unfortunately, as a defining criterion laughter is both too wide-ranging and too narrow. In certain cultures, for example, laughter is a sign of embarrassment, while even within British culture laughter may not be connected with humour and may systematically evoke a serious response (Jefferson, 1984) or be designed for other 'serious' interactional purposes (Jefferson, Sacks and Schegloff, 1987). Attardo, in a recent discussion of this topic, concludes that research evidence 'has established beyond doubt that humor and laughter, while obviously related, are by no means coextensive' (2003: 1288), a position that captures the essential relationship and undermines the definitional credentials of laughter.

In the analyses that follow I adopt the very broad definition of humour offered by Long and Graessner (1988: 37): 'Humour is anything done or said, purposely or inadvertently, that is found to be comical or amusing.' Within this, laughter will be treated as an indication of humour but no more, an example of the 'tying' that Sacks identified in the relationship between laughter and that to which it is designed to respond (1992b: 74). Humour, as Tannen observes, is perhaps '[o]ne of the most highly distinctive aspects of any person's style' (1984:130), and the relevance of individual style will be considered, though the focus will be on general features. In all cases careful consideration will be given to the structuring, sequential positioning and content of an utterance, as well as to other factors indicating humorous design, such as the presence of what is known as a *candidate laughable* (e.g. a pun or the punchline to a joke). Analytically, my position corresponds closely to that adopted by Holmes:

> Instances of humour in this analysis are utterances which are identified by the analyst, on the basis of paralinguistic, prosodic, and discoursal clues, as intended by the speaker(s) to be amusing and perceived to be amusing by at least some participants
>
> (Holmes, 2000b: 163)

In terms of transcription my approach is more delicate than that of most approaches (e.g. Holmes and Marra, 2002), but less refined than the norm for CA because of my occasional use of italics to indicate utterances delivered with a laughing voice in preference to drawing on Jefferson's very detailed system for capturing laughter (1984, 1985). It is possible that as a result I have missed some subtler points relating to the precise timing of the onset of laughter or the nature of its delivery, but none that would undermine any of the claims made in what follows.

The power to involve

> [Humour] provides, at least temporarily, a group unity or awareness, a psychic connection of all the laughers. It can be induced as a means of displaying this group togetherness. It allows for the expression and maintenance of group values and standards, via the subjects and situations to which it refers. It can boost morale and ease internal hostilities and differences. Laughing at people or things external to the group can strengthen boundaries, solidifying members in their group identity against outsiders. The importance of shared laughter is lost in studies that treat laughing exclusively or primarily as the product of the individual mind, or which rely on methods that test the laughing response in isolated individual subjects.
>
> (Glenn, 2003: 30, drawing on Hertzler, 1970: 93–8)

By its very nature the world of humour is one of inclusion and exclusion: those who 'get' the joke join in the laughter it generates, which in turn excludes those who fail to understand or appreciate the humour involved – or, more poignantly, those against whom the humour is directed. In the case of a collaborative group, especially one where the avoidance of conflict has a high priority, such exclusion will be avoided wherever possible and humour will be exploited as a means of reinforcing group solidarity. 'Joking', as Norrick observes (2003: 1342), 'works to establish and enhance group cohesion, and serves as a control on what sorts of talk and behaviour are acceptable to participants in the interaction.'

The control function of humour has long been recognised and is usually paired with a complementary dimension labelled 'conflict' (Stephenson, 1951), 'resistance' (Powell, 1988), 'social tension' (Paton, 1992) or the like. These reflect an essential distinction between, on the one hand, the expression of common sentiments, perspectives, judgements and the reduction of tension or awkwardness and, on the other, the expression of hostility or aggression. More recent studies, however, have tended to concentrate on the former dimension, emphasising the importance of 'doing collegiality through humour' in the workplace (Holmes and Marra, 2002: 1686) and reinforcing group solidarity. Zadjman, for example, has argued that 'constructive humour' serves not only to lubricate social relationships but also to maintain the positive face of the group through fostering consensus and solidarity and strengthening personal bonds (1995: 327).

In this chapter I explore the dimensions of constructive humour and reveal the different forms it takes, arguing that the nature of such humour is largely determined by the relationship between the relevant setting and professional engagement. This represents an attempt to establish some parameters of the relationship between humour and 'business' in a way that has not to my knowledge been attempted before and in so doing enables distinctions to be drawn between interactional practices that, for perfectly understandable reasons, are common across professional work groups and those practices developed by particular groups over time through repeated instantiations.

The analyses that appear in the following sections, then, examine three different relationships between professional business and the possibility of workplace humour:

1. Situations in which humour does not occur.
These are situations where *as a matter of fact* humour is not to be found. They are therefore not the sort of situations that Woods (1979) has in mind when he talks of laughter sometimes being excluded by 'laughter inhibitors'. There are no detectable laughter inhibitors in these exchanges; they are simply constructed in a way that leaves little if any space for humour. Neither are they explicitly signalled as 'serious business' in the way that certain exchanges are in the Pen staff room.

2. Events in which humour occurs and takes much the same form.
These examples feature prominently in mainstream workplace humour research. Although the forms of humour may differ slightly, its function

is fairly constant and is to be found in all three settings. The examples here, as with others in the literature, are all taken from meetings.

3. *Places where humour is to be found but where it takes very different forms.* The nature of these differences, I will suggest, derives from the relationship between the setting and the institutional world of which it is a part. The final section in this chapter argues that different collaborative groups may have very different ways of doing humour.

Identifying humour is one thing, but establishing patterns across cases is altogether more challenging. In fact, some researchers, while recognising that differences exist, are wary of attempts to pin them down in terms of particular features. In the introduction to his summary of research into humour in the workplace, for example, Paton warns the reader that 'no clear pattern of joking relationships and techniques emerges from the studies conducted to date, but a range of "unofficial conventions"' (1992: 2). However, this might simply reflect individual differences rather than interactional norms associated with particular groups. Humour covers a lot of territory and it would be surprising if everyone in a particular group shared exactly the same sense of humour and a predilection for the same style of humorous interplay. As an illustration of this I take an example from the Pen staffroom. The following exchanges occur within about five minutes of one another and demonstrate Harry's preference for humorously completing other people's utterances, a form of joking not used by any of his colleagues:

Extract 4.1

```
001   Paul:  'I was a imme- no sooner had I walked into
002          the room than I was immediately struck by'
003 Harry:  Hheheh a brick.
```

Extract 4.2

```
001   Paul:  Yeah. An::d (.) sh- (.) 'Ye:s but
002          (0.5)
003   Paul:  I:
004          (0.5)
005   Paul:  it's difficult for me to:'
006          (0.5)
007   Paul:  and then she didn't
008          ┌finish the sentence.
009 Harry:  └'Express myself.' Heheheh
```

As will become apparent from the following analysis, this style of humour neither reinforces nor works against the dominant patterns that help define this group, although a study of individual style might itself prove interesting (the laughter that accompanies the completion in the extracts above, for example, is produced by the speaker and not reciprocated). A distinction perhaps needs to be drawn between styles and patterns (or norms) of humour, where the former could be assigned to either individuals or groups, while the latter would be jointly established and therefore associated only with groups. Style is not necessarily neutral, and it would be possible for someone's style to work against the group's ways of constructing humorous talk, but the response of other members of the group to this would be marked. It is certainly unlikely that such a position could long be sustained within a collaborative group any more than Helen was able to maintain her role as Chair in the face of group resistance. In what follows, then, individual style will not be analysed except where particular occurrences are relevant to the wider analysis.

For the same reason, no attempt will be made to categorise types of jokes and present a numerical analysis of these for the purposes of analysis, except for a single instance where this serves to provide a useful brief overview of humour in the Pen staff room, and even here this is not used as part of the analysis. The temptation to compare humour in different settings in this way is a natural one, but because it cannot hope to capture the subtleties of ways in which humour is worked for and responded to in different instances it might easily be misleading. So although examples may be grouped together where there are similarities in terms of topics or approaches, this should not be regarded as a reflection of systemic patterns but rather as an organisational convenience that allows examples to be compared. As Glenn has observed (2003: 38), '[c]laims of a recurrent pattern do not rest on frequency or statistical probability but on demonstration in examples and explication of the practices, orientations, rules, competencies, expectations, etc., which participants display in their conduct.' The humour examined here should be seen as part of a much bigger picture, as its reappearance in subsequent chapters will confirm.

Doing business together

One of the interesting features of the data set for these three groups is the extent to which humour is peppered throughout their talk. Sometimes it appears in the form of asides and individual exchanges and sometimes as a shared enterprise constructed over a number of turns, but it is never far from the surface. However, there is a particular form of talk where it is not to be found, and it is interesting to consider why this should be so. Perhaps even more interesting is the fact that the features of this talk are to be found across all three groups as the following three examples reveal. All are taken from meetings and all could be labelled in the same way. I present them as a group in order to expose more starkly their characteristic features:

Extract 4.3

Pen pre-course preparation meeting
Having discussed with the group the timetable for the coming term, Jenny, the principal, introduces the subject of rooms.

```
001    Jenny:   E:m (.) and the other thing (.) we haven't
002             decided is rooms (.) that people should be
003             in.
004             (0.5)
005    Jenny:   This course. Whether we want to change,
006             (1.0)
007    Jenny:   stay the same, (.) or ┌I
008    Paul:                          └I (xxxxxxxxx) could
009             stay the same.
010  Annette:   Yes I might have to change depending
011             on ┌numbers.
012    Jenny:      └size.
013    Jenny:   Mmm=
014  Annette:   =So perhaps tomorrow I'll start in (.)
015             seven, continue in seven,
016    Jenny:   Mmm
017  Annette:   and then depending on (.) how many I've
018             got. Extra.
019             (0.5)
020  Annette:   Em (.) where else is available? Th- you're
021             in: er
022     Paul:   I'm in one.
023  Annette:   One.
024     Paul:   °Yeah.°
025  Annette:   An:d you're in
026       Ed:   Five.
```

```
027  Annette:  Five. Right. ⌜So I can move⌝ up to two
028   Jenny:            ⌞(xxxxxxxxxxx)⌟
029            (0.5)
030  Annette:  ⌜⌜I'll move up to two if           ⌝
031    Paul:   ⌞⌞I would just like to keep⌟ the continuity
032            for ⌜the pe⌝ ople who are there,
033   Jenny:      ⌞Mmmm ⌟
034    Paul:   I mean ⌜if ther⌝ e's I mean=
035  Annette:         ⌞Mmmm ⌟
036    Paul:   =if ⌜there's⌝ a crisis=
037   Jenny:      ⌞ Mmmm ⌟
038    Paul:   =then obviously I'll move out and
039  Annette:  No it's- that's fine because I'll start in
040            seven, and if necessary I'll move to two.
```

Extract 4.4

DOTS workplan meeting
Sue has been working through the timetable, entering
'out, in, out, in...' and calling these as she does so.

```
001    Sue:  The second wee:k.
002          (0.6)
003    Sue:  Tuesday, will be:
004          (1.0)
005    Sue:  a seventy two hour out, °seventy-
006          (0.2)
007    Sue:  two°
008          (0.4)
009    Sue:  ⌜⌜ou:⌝
010  Penny:  ⌞⌞out⌟
011          (1.0)
012    Sue:  Put you::r
013          (2.0)
014    Sue:  E::⌜:::m::
015  Penny:    ⌞(xxxxx at a ti::me)
016          (2.0)
017    Sue:  Bring it a:ll °(see) two hour out.° Miss out
018          your twenty four,=°put° two hour in,
019    Sue:  °two (.) in,°
020          (1.0)
```

```
021   Sue:   [[Which has come ⌐ out (on) Tuesday,
022   Penny: [[(xxxxxxxxx out)⌐
023          (1.0)
024   Sue:   so you're on schedule for this then,
025          (1.0)
026   Penny: Yeah, but I've still got my
027          (0.6)
028   Sue:   So on ⌐Fri:da:y ⌐ you put your six hou:r
029   Penny:       [(xxxtable)⌐
030          (0.2)
031   Sue:   in, >you take< your six hour (.) ou:t,
032          (5.0)
033   Penny: Yeah. The only thing is then,
034          (1.0)
035   Penny: I >want them< bo- I want each timetable to go
036          in: at the same time of day >if you see what
037          I< mea:n.
038   Sue:   No it will.=But you just swap (it) round like
039          you ⌐take⌐ your twenty four hour bags
040   Penny:     [mm ⌐
041          (0.6)
042   Sue:   out of there, move 'em from the other day, and
043          put your twenty four hours in
044          (0.4)
045   Penny: at the end.
046   Sue:   another da:y. That works it so you're not
047          doing Fri- (.) overtime
048          Friday ni ⌐ght.
049   Penny:           [>Friday night.<
```

Extract 4.5

IEC weekly meeting
Having discussed procedures for extensions, the talk
has moved on to the issue of the new Diploma that can
be awarded on completion of 100 credits and a Diploma
Portfolio worth a further 20 credits. Helen has just
clarified a point relating to this.

```
001   Paul:  In th ⌐is CAs ⌐ :e
002   Helen:      [Right ⌐
```

```
003              (0.4)
004    Paul:     is i- (.) would it be possible for the
005              per-=for us to give the- the person the
006              adVIce to sa:y
007              (0.6)
008    Paul:     IF you are DEFinite about going on the MSc
009              then it's in your wh- while to do · h hundred
010              and twenty-
011              (0.2)
012    Paul:     bona fide-
013              (0.2)
014    Jack:     ↓ YES:. Y ┌es.
015    Paul:            └CREdits (.) and
016              n┌ot waste your time on °the°┐
017    Helen:    └Yes. And not waste your ti ┘me on that.=
018    Paul:     =po┌rtfo┐ lio.=
019    Tony:        └Yes.┘
020    Helen:    =Tha┌t's (xxxxx) yeah.┐
021    Tony:         └'nd AT which poi: ┘nt,=
022    Helen:    Coo=I >knew we'd get good ideas< °
023              (if ┌xxxxxx┐ xxxx)°┐
024    Paul:         │°yeah°│       │
025    Tony:         └At whi┘ ch poi┘nt we have also
026              agreed,
027              (0.8)
028    Tony:     I ho┌pe,
029    James:        └((clears throat))=
030    Tony:     =E::m,
031              (0.6)
032   James?:    (x)hh ┌hh
033    Jack:           └I think
034              we H ┌ A:VE.=I-I th ┐ ink ┌wehave.┐
035    James:         └((clears throat))┘      │        │
036    Tony:                              └(At that┘
037    Tony:     (xxxx) but we could issue a letter,
038    Jack:     uh
039              (0.4)
040    Jack:     u ┌:h
041    Paul:       └Mmmm ┌mm
042    Tony:            └>saying-<
043              (0.2)
```

```
044  Tony:  'To whom it may conce::rn,
045         (0.4)
046  Jack:  Y es.
047  Tony:   ay en other has no:w:
048         (0.2)
049  Tony:  having- >been registered on< our:
050         (0.6)
051  Tony:  D(x)iploma=>MSc Pro<gramme,
052  Paul:  Mmmm=
053  Tony:  =has now got enough credits to be aWArded the
054         Diploma.
055  Jack:  Yes.
```

These passages, all of roughly the same length, exhibit a number of identical features apart from the absence of humour. A full analysis would be inappropriate in a chapter on the subject of humour, but the following are indicative:

- They are all concerned with arrangements. The Pen focuses on room assignments, in DOTS the concern is with a schedule for procedures relating to an experiment currently under way, and IEC colleagues are working out new procedures in response to a change in the rules for a course they teach.
- Even allowing for the preference for agreement which obtains in talk (Sacks, 1987), the frequency of explicit agreement is particularly noticeable and on the only two occasions where disagreement occurs 4.3:39 and 4.4:38 it is actually supportive of the prior speaker as a colleague.
- There is quite a lot of overlapping talk and perhaps the most interesting aspect of this is the evidence of joint construction, where the hearer projects the end of the prior speaker's turn and provides a completion to it, sometimes in the same form as the first speaker ('out' in 4.4:9–10) and sometimes in a different form ('size' and 'numbers' in 4.3:11–12; 'at the end' and 'another day' in 4.4:45–6). The most emphatic occurrence of this is in 4.5:16–18, where Paul is completing his statement to the student, which Helen echoes, providing the projected completion 'that'. Paul allows her to do this and latches his own conclusion onto the end of her turn. Helen's turn-initial agreement ('Yes') has been emphatically ratified by both participants in a jointly constructed conclusion to his statement.

- This exchange also features repetition, which is to be found in the other samples as part of the process of agreeing or confirming arrangements. In 4.3:1–11, for example, when Jenny introduces the topic of rooms, the arrangements offer two possibilities, 'stay the same' or 'change', which are picked up in turn by Paul and Annette in the articulation of their responses. At the end of 4.4:48 Penny's slightly overlapped reiteration of Sue's 'Friday night' confirms the latter's assessment.
- Co-operative overlap and participatory listenership are aspects of Tannen's 'high-involvement' style (1984), reflecting the participants' professional orientation to the topic. However, the strategic placement of overlap towards the end of the speaker's main point allows sufficient space for the speaker to complete this without interruption. Hence the silences, some quite long, within the development of a point (e.g. Jenny in 4.3:1–6, Sue in 4.4:1–9 and Paul in 4.5:1–13). The whole of the IEC extract reflects this organisation. First Paul develops his position without interruption up to line 14, where Jack emphatically agrees with him. It is worth noting that this agreement comes after a short pause following 'bona fide', when it is possible to predict 'credits', which in line 15 Paul uses to complete his main point. Following confirmation that Jack, Helen and Tony are in agreement with this and having established his right to the floor, Tony then develops the text of a letter, prompting only continuers from Jack and Paul and final confirmation from Jack.

Exchanges of this sort are common in the data and one of the features which allow them to be identified is the absence of humour. There are certainly gaps where humour might arise in the above extracts, and it might also be argued that the jointly constructed positions and shared agreement establish ideal interpersonal conditions for its introduction. However, in all these cases, a carefully articulated position is being developed, and this calls for a response from other participants in order for agreement on arrangements to be reached. It is also important that the nature of that agreement should be clear to all. Humour has many functions, and it may serve to make particular occasions or occurrences memorable, but from the perspective of establishing clear points that everyone understands and all are expected to remember (or at least note) it represents a potential distraction. Therefore, when the topic involves making arrangements, agreeing procedures or similar business, the talk focuses on matters of detail and remains essentially serious; involvement is generated by other means such as overlap and repetition.

Meeting relief

One of the most general contributions humour can make to professional talk is that it 'escapes the straitjacket of business talk' (Holmes and Marra, 2002: 1697), and nowhere is this more true than at meetings. By their very nature these are concerned with matters of business, and their success depends on treating this seriously, so they tend to be found at the same end of the entertainment scale as picking oakum and counting sheep. In such a context humour provides a welcome form of relief, though its functions are not necessarily confined to this and its forms vary.

Because meetings, despite their general unpopularity, are recognised as professionally important, occurrences of humour are rarely extended. In fact, there is only one example of this in my data set, notable because of the unusual circumstances in which it occurs. The relevant agenda item appears as 'answer machine', but the person who had asked for it to be included is not present and nobody at the meeting can suggest a reason for it since a new machine has just been bought and is functioning perfectly well. This is therefore a redundant item and the ensuing 150 lines, which comprise humorous exchanges on the subject of answerphone messages, are effectively a replacement for the slot that it would otherwise have filled. The fact that this is also the last item on the preliminary agenda before the main item is introduced also means that in some respects it corresponds to pre-meeting talk, which is often characterised by humour (Holmes 2000b: 179). Finally, and in some respects most importantly, it allows the Pen staff to indulge in the sort of shared exchanges that characterise so much of their humour, as the next section and next chapter will illustrate.

Humour in meetings, then, tends to be in some way related to the topic under discussion, brief and relatively mild – in fact, minimally intrusive. These characteristics perhaps explain why meeting humour, whatever its distinctive patterns, tends to arise from similar sources across settings. The following topics are not exhaustive but they are the most prominent and are to be found in all three sites.

'Playing off' the meeting

In view of the importance of covering the relevant business during meetings, it is hardly surprising that one common form of humour involves playing off aspects of the meeting itself, using serious responses on an agenda item as a basis for mild amusement of the sort that does not interfere with coverage of the agenda itself. The best example of this comes from a Pen meeting where the addition of mildly amusing modifications to proposals develops as a theme that is picked up in different agenda items at different points in the meeting.

In the first extract the Pen teachers are discussing the establishment of a new system for pooling teaching ideas and resources, which provides an opening for a comment highlighting the tendency of systems to generate the need for other systems to support them:

Extract 4.6

```
001  Annette:  E:m (.) well we could have that as a pending
002            box. Before you put something in the::re
003            while er well before we had the staff
004            meeting on it,
005     Paul:  Mmm
006  Annette:  but (xxxxxxxxx)
007     Paul:  I (feel) we should have another box.
008            (0.6)
009     Paul:  A pre-pending=
010  Annette:       ⌈⌈Hahahahah
011    Keith:  =⎢⎢Heheheh
012    Harry:       ⌊⌊Hahah
```

The humour here, which prompts brief laughter from those present, highlights the oddness of having a box in which papers or materials relevant to a forthcoming staff meeting might be deposited. Rather than challenging the practicality of this, which would certainly have taken longer and might have prompted an extended discussion, Paul undermines the proposal by suggesting a further box to perform the same function, this time for the pending box itself. His delivery of the joke is well worked, allowing the apparently serious proposal in line 7 to stand for over half a second before delivering the punchline which undermines it.

In the second example from slightly later in the same meeting, the staff are discussing the establishment of a 'teaching room for dedicated business English teaching' and have agreed that this would be feasible only at certain times of the year:

Extract 4.7

```
001     Paul:  But if it's dedicated then (.)
002    Harry:  Hhhhh ⌈YEah.
003  Annette:        ⌊Yes
004    Harry:  >Sort of semi< dedicated.
005  Annette:  Yeah::hahahah
```

Linguistically, this joke is identical to Paul's earlier one, but its pragmatic force is weaker since it does not undermine the proposal itself. It is also more of an aside than a worked-for joke in the same style as Paul's, and perhaps as a result only one colleague laughs. Although this is little more than an aside, it nevertheless indicates that staff do orient to particular styles of joking, establishing patterns that may be local to particular meetings or, if repeated over time, characteristic of the group's interactional style.

In the final example, taken from much later in the meeting, teachers are discussing the post of academic manager, which disappeared in the cuts of a couple of years before and has been sadly missed since, cropping up regularly as a source of concern and comment:

Extract 4.8

```
001  Annette:  Well I think we ought to look
002            at the section >because< the first
003            one ┌on that, the: academic manager,
004  Harry:       └Yes.
005  Harry:  Absolute ┌(xxxxxxxxxxxxxxxxx)┐
006 Annette:          └I think we must (.)┘absolutely
007          must have one.
008    Paul:  °Absolutely agree.°
009 Annette:  >Plus it must be a< permanent part time
010           position.
011  Harry:  Yeah.
012          (1.5)
013  Harry:  Permanently rotating.
014    Paul:  HEHE ┌HEhahahah┐
015 Annette:       └Hahahahah┘ HAhah=
016   Harry:  =With a turntable in the middle of the
017           floor, and · hhh hh
```

Harry's play on words here, made explicit in his follow-up (l.16) to the original joke, depends on the shared assumption that the position will be a rotating one as in the past. This enables him to pick up on the juxtaposition of 'permanent' and 'part time', adding the idea of rotation and casting the whole in a humorous light. As with the other examples here, this joke enables the group to establish a degree of humorous but at the same time critical detachment from the systems they work within but also seek to refine. As such, humour seems to play a subtle but important role in the developmental dimension of a collaborative culture: by making systematisation relevantly funny, the group expose it to a scrutiny that serves as a useful mechanism for exposing distortions of expectation or assumption. The danger of collaborative interaction lies in the power hidden assumptions can exercise on the decision-making process, and for this reason it is important to create suitable distancing mechanisms in order to establish a critical perspective on group decision-making. The ability to establish detachment through humour, which thrives on incongruity, ambiguity and inconsistency, is therefore an important contributor to the group's own sense of balance.

Linguistic opportunism

Of course, not all joking functions in this way; sometimes it amounts to no more than seizing an opportunity to be funny. The quickest and easiest way to achieve this – and also the least time-consuming – is to exploit a linguistic opportunity. Such humour takes various forms, although puns are relatively rare and linguistic humour generally seems to be more preferred by the Pen teachers than by the other two groups. All three groups, however, seem open to the potential of suggestive ambiguity for humorous exploitation, as the following examples illustrate.

Of all the groups, DOTS is by far the readiest to seize on opportunities for sexual innuendo and, outside formal meetings, to make sexual activities and proclivities a butt for humour. The following example concerns bodily rather than explicitly sexual functions but is interesting because it derives from a perfectly innocuous term which might easily have been missed by a group less concerned in their daily work with bodily processes. The meeting has just been started by Julie who is now reading the relevant regulations for record keeping from an official document:

Extract 4.9

```
001      Julie:    ↑TRAIning records
002                (2.5)
003      Julie:    reco:rd reTEntion,
004                (0.8)
005      Penny:    T!hh°hh°h
006                (0.8)
007      Penny:    HHh!HwhhhO ┌OHH!
008   Margaret:               └HE ┌HhehEHEHeh┐ eh=
009      Julie:                   └Hahahahah  ┘
010   Margaret:    =heh ┌eheheHHE!°he┐ heheheh°
011      Penny:         └Get it out!  ┘
```

This illustrates the importance of the placement (Jefferson, 1979) of the first laugh, which displays to the other participants that the laugher takes its referent (in this case a prior turn) to be funny. There is nothing funny about 'retention' until Penny makes it so. Certainly, there is no indication in Julie's utterance of the term or, given the regularity of long pauses in her presentation so far, in her pause after it to suggest that this is offered as a candidate laughable, and Penny's first response (l.5) is very brief. The others join in only when she laughs again after a pause in which she seems to suppress the urge to do so, prompting her in line 11 to make explicit where the joke lies.

By comparison, the Pen example is relatively mild and depends on explicit signalling. Again, the team are working their way through a list being read out, this time from a pro forma for the report that has to be submitted by the school in preparation for an inspection visit. A number of points have already been covered:

Extract 4.10

```
001      Jenny:   'External positions held by staff.' Got a
002               funny heading here.
003               (0.8)
004      Jenny:   Got (.) '(xxxxx) ⌈job'
005      Annette:              ⌊Ex(.)ternal position.=
006      Jenny:   =Yehe:s. Heh ⌈that's right.  ⌉
007      Louise:               ⌊(xxxxxxxxxx)   ⌋
008      Jenny:        ⌈⌈Yeaheheh ⌉
009      Annnette:     ⌊⌊Hahahah ⌋ I see what it ⌈it means now.
010      Jenny:                                 ⌊ Right ⌈(xxxxx)⌉
011      Kate:                                          ⌊(xxxxx)⌋
012               seventy third, yeah.
013      Jenny:   Yehes.=
014      Paul:    =Heheh. Heh.
015      Kate:    Oh God. (That's been recorded.)
016      Jenny:   He ⌈heh ⌉
017      Paul:       ⌊Heh⌋eheh
```

Having drawn attention to the ambiguity of 'external positions', Jenny leaves a pause, but when no laughter is forthcoming she continues with her main point. Only when Annette repeats the term does she indicate that her reference was intended to be humorous (l.6), prompting shared laughter (ll.8–9), Annette accounting for her initial failure to laugh by explaining that she had failed to understand. Kate then alludes to the Kama Sutra in her use of the ordinal number, and this prompts laughter from Jenny and Paul, which Kate manages to extend by referring to the presence of the tape recorder. The joking has been mild enough, but it provides useful relief from the relative tedium of working through points on a form.

As in the case of DOTS, in IEC it is an individual who picks up on the humorous potential of an expression, though in this case it is an invented expression. The team have spent some time discussing the complex issues associated with graduating from the Master's programme with a Diploma then returning to it at a later stage and Jack has just invited Tony to introduce the next topic on the agenda. However, when Helen asks who is going to be responsible for noting down all the relevant procedures and regulations, he responds supportively:

Extract 4.11

```
001   Jack:   I:'ll send you my I'll send you my:
002           (0.2)
003   Jack:   diddlyflop,
004           (0.4)
005   Jack:   my er- ┌the- (.) o┐ ffi┌cial          ┐
006   Kathy:         └ °A::hhh° ┘    │              │
007   Sally:              └ t!HHHHHhh ┘eh┌hehheh ! ┐
008   Jack:                             └diddlyfl┘op
009           that is, =NOT NO=
010   Helen:  =I don't know >whether I want< your
011           ┌ diddlyflop (y'can keep it.)        ┐
012   Jack:   └ NOHHHHHHAHHAH!=Well y'r 'av┘in' ┌it.┐
013   Julie:                                   └°hu┘ hheh=
014           hehe ┌ha:h°┐
015   Paul:        └°u:mm┘ m°=
016   Jack:   =I'll email it so it'll be perfectly ha:rmless.
017   Sally:  ((Squeak))┌=hehHEH!   ┐
018   Paul:            └°Tom Book┘(in xxxx)°
019   Jack:   There'll be no VIrus involved.
020   Kathy:  Huh!he::hh!
```

Jack's use of 'diddlyflop' seems to be a genuine case of inserting an invented term after an unsuccessful wordsearch. The lengthened 'my' followed by a brief pause invites completion, and the term is used when this is not forthcoming. At first it appears that Jack wishes to continue as normal and there is even some evidence (l.9) that he continues with this line after Sally's laughter. However, when Helen herself picks up on the ambiguity and orients to the impropriety by repeating it (l.11) in the next turn (Jefferson et al., 1987: 160), he laughs and provides a teasing response, which in turn produces laughter from Julie. The sexual dimension of the joke is then pursued through the idea that if he emails his response this will ensure that it is free of viruses. At the end of this extended joke, Sally, Jack, Helen, Julie, Paul and Kathy have all joined in either with laughter or comments, though the nature of Paul's contribution, referring to a university IT specialist (l.18) and uttered while laughing, is not clear from the transcript. They have also contributed to the development of the joke because of the three ways to extend shared laughter identified by Glenn (2003: 73), only 'extend the laugher without more laughables' is not exploited: Helen (l.11) re-invokes the first laughable and both Jack (ll.16 and l.19) and possibly Paul (l.18) provide nexts in a series of laughables.

This is very much a joint enterprise in which all have participated, yet there is no overlapping laughter and it seems that different individuals find different aspects funny. Perhaps this is a feature characteristic of humour that serves only as a form of immediate relief, arising from its tendency to impact on the individual rather than the group. The affiliative functions of humour are particularly important in collaborative groups, but its spectrum is much wider than this. As Woods notes (1979: 215), humour 'can be intrinsically satisfying, over and above the instrumental gains that might be got from it and indulged in for sheer delight', and if the laughter is not always shared, this still allows individuals to find mild amusement in a way that suits them.

Teasing

Teasing is an altogether different proposition. However intimate the target, jokes against the person, especially if delivered in the company of others, are highly face-threatening and have to be handled with considerable sensitivity if interpersonal relationships are not to be damaged (for a discussion of relevant face issues, see Zadjman, 1995; Holmes, 2000b) Outside meetings there are considerable differences among the three groups, with teasing very common in the DOTS group, evident in the IEC setting, but rare in the Pen. Within meetings, though, there seems to be little difference in the distribution, perhaps because there is less latitude here for such humour. Teasing takes up time, is personal rather than general, and is very unlikely to be related to the business in hand. It is therefore relatively rare.

When it does occur, however, it is rendered innocuous by the simple device of exploiting accepted stereotypes and/or tapping into running jokes. The secretary of the Pen, for example, is held up as the authority figure in the school and in this guise provides her colleagues with one of their running jokes, as the following entry from my fieldnotes demonstrates:

> Along the same lines, though falling neatly into the 'powerful secretary' bracket, was another incident. During the first phase of the morning Pat had asked me to make a few copies of a student information handout and in withdrawing the first sheet from the copier I'd pulled off a map of Quillham. It was very loosely glued and the damage was therefore negligible, so I was able to make a joke about it when I returned it to Pat. Just before I left she popped into SR1 [the staff room] and I jokingly apologised for not actually doing the post office run as I'd promised when I arrived. This led on naturally to my shamefaced admission that with that and the photocopying failure I'd got off to a rotten start, and the general lighthearted confirmation that I must be in Pat's bad books. Jenny's contribution was, 'Now you're in the same boat as the rest of us.'

Pat's role as someone not to be tangled with is also used as the basis for the following teasing at Harry and Pat's expense during a morning break. Harry is acting academic manager, so the appearance of Pat's black tights signals 'business':

Extract 4.12

```
001  Annette:  Yes it's (.) as soon as those black
002            tights appear round the corner Harry (.)
003            hehe ⌈ heheh ⌉
004   Harry:        ⌊ Heheh ⌋
005            (0.5)
006  Annette:  gets ⌈ worried. ⌉
007   Paul:         ⌊ Well she ⌋(xx ⌈xxxxxxxxx)
008                                 ⌊((General laughter))
```

It is this background that allows Paul to tease Pat, who is secure in her reputation, without risk of giving offence. The meeting is discussing the appointment of an office junior to support Pat and it begins with an exchange that can only serve to remind colleagues that her reputation is well-earned:

Extract 4.13

```
001  Annette:  And the next office junior presumably::
002            (1.0)
003  Annette:  you mentioned that (.) did you?
004            (1.0)
005     Pat:   You should know. You were there.
006   Harry:   hhh hh heh!
007     Pat:   You did my appraisal.
008   Harry:   Heh yes
009  Annette:  ⌈Hehehehehehheh             ⌉
010    Paul:  ⌊I-I-I- (accept) the-⌋(.) I
011            don't ⌈think⌉ we=
012  Annette:        ⌊But-  ⌋
013            =should ⌈really be referr⌉ing to Pat as=
014  Annette:         ⌊Don't think I  ⌋
015    Paul:   =the office junior actually.
016            ((General laughter))
```

Referring to the proposed appointment, Annette seeks to confirm that Pat has stated her need for this support and receives a response from Pat (l.5) implying either that the question is redundant because Annette already knows the answer or that the latter had failed to pay proper attention during the appraisal she had conducted. This very forthright reply in a style that is by now expected produces laughter from Harry and, subsequent to this, from Annette, who then tries to develop her point. However, Paul introduces and persists with a joke that depends on a deliberate misunderstanding exploiting the incongruity of Pat's robust response to Annette and the latter's purported description of her as 'office junior'.

This exploitation of a running joke may become such a standard routine that the laughter is drained from it, the participants effectively assuming standard roles with which they are already well familiar. The following extract from IEC dates back to an occasion when Jack had taken the external examiner for the Master's programme out for a meal at a nearby restaurant. This is standard practice and all would have gone well had Jack not mistakenly entered 2.30pm rather than 2pm in his diary as the time of the examination board. Familiar with his normal punctuality, his secretary telephoned the restaurant when he did not

appear at 2pm, but payment of the bill and the walk back meant that
the board was not able to begin until 2.15. Although everyone was
understanding at the time, this subsequently became a source of teasing,
even exploited on occasions by Jack himself. In the following extract
from a meeting to confirm arrangements for another board, his reference
to just being able to sit back and let things take their course provides
Tony with an opportunity too tempting to ignore:

Extract 4.14

```
001   Jack:    If the other stuff's as good as that we can just sit
002            back and °let ┌ it°
003   Paul?:              │ Well (if that)┐
004   Tony:              └ The good thing┘ about that
005            arrangement as well > that ┌ it-<┐ it=
006   James?:                           └ w-  ┘
007            =sa: ┌ ves  the  rest ┐ of us just sittin' rou:nd=
008   Helen:       └ Yeah=we go' a- ┘
009   Tony:    =while you're maundering off having lunch some
010            ┌ where.
011   Jack:    └ That's ABso ┌ lutely  ri ┌ ght.
012   Kathy:                └ (We:ll you │ xxxxxx xxxxx me: )┐
013   Sally:                             └ And forgetting wh┘ at
014            time to ┌ come back.┐
015   Kathy:           └ And forge ┘ tting what time to
016            come ↓ back.
017   Tony:    MMMM!
018   Kathy:   Yes.
019            (1.0)
020   Jack:    O:N: ┌ :EMISTAKE!=>(I've had this)<O:N:::: ┐ E=
021   Kathy:       └ And comPLAINs that ┌ he's (mak ┐ ing-)┘
022   Helen:                           └ Why don't-┘
023   Jack:    =mis ┌ t↑a:ke >'n then (it goes xxxxxxx xxxxxxx)┐
024   Helen:       └ Why don't you have the ┌ exam board fi:rst,│
025   Kathy:                               └ He doesn't complai┘:n
026   Kathy:   =th ┌ at   he ┐ doesn't=
027   Jack:       └ (LIfe!)┘
028   Kathy:   =give it- ┌ (the xxxxx xxxxxx) ┐
029   Helen:            └ Why ┌ don't ┐ you ju ┘st ┌ take it the: ┐ re.
030   Tony:                   └ OH! ┘              └ I FORGOT. ┘
031   Sally:   (x)HE! ┌ heh!
032   Tony:          └ The tape recorder's ON.
033            ((General laughter))
```

This is a story to which all can contribute: Tony presents his mock complaint about Jack 'maundering off having lunch somewhere' to which Sally adds 'and forgetting to come back', repeated by Kathy. Meanwhile Jack exaggeratedly bemoans what has befallen him for just 'one mistake' and Kathy tries to develop a further point about complaining, while Helen offers suggestions for avoiding a repeat of the incident: have the exam board first or take it to the restaurant. Having remained silent as these actors contribute to the shared performance, Tony now returns with a fresh angle on an old story. In raising the issue of Jack's indifference to the fate of the exam board and his role in it, he has 'forgotten' that the tape recorder is running: the embarrassing incident of the belated exam board is now a matter of public record. This new twist earns the general laughter it deserves.

'Through laughing, and laughing together,' says Glenn (2003: 2), 'we contribute to the ongoing creation, maintenance, and termination of interpersonal relationships.' The imperative in the three groups, all working within narrow physical and professional territory, is to achieve the first two of these and avoid the last at all costs. The method they employ to do this where teasing is involved is to follow well-trodden paths, and preferably those already sanctioned by the recipient.

In-jokes

Eventually, as in the case of the last example, routines like this become in-jokes, the humour entirely lost on an outsider and the steps so well worn that they lose the edge that sparks laughter. Such internal humour by its very nature reinforces group bonds, but a wider professional context is also available as a resource for in-jokes, allowing the possibility of strengthening group ties through the recognition of shared professional knowledge and experience. The following joke is an example of this. It occurs in a staff meeting given over to a joint report by Jenny and Harry on a professional presentation they had attended at a teachers' club. The topic is the teaching of pronunciation and Harry is running through a series of options for teaching particular sounds:

Extract 4.15

```
001  Harry:  And had it sorted E:::m, which one's that?
002          Learner speaks and teacher (.) points.
003          (1.0)
```

```
004    Harry:   E:r, learner points and learner speaks, is
005             the next.
006             (2.5)
007    Harry:   If I've got it right In class you've got
008             to say whether they're right or not, I
009             suppose.
010    Jenny:   Mm
011    Harry:   This is what I think I'm saying, is- is that
012             right?
013    Jenny:   (Also xxxxx) have differences between
014             the learners themselves. (The) French
015             and the Spanish and the Japanese all (.)
016             having got rather different (.)
017             er approaches to (learning to)
018             produce a sound.
019    Ed:      Hah
020    Harry:   Mm
021    Jenny:   And they can start to home in on it and can
022             identify how- the variety of this sound
023             tends to feel. (That they produce) can be:.
024  Annette:   Uhhuh
025             (1.5)
026    Harry:   Mm
027             (2.5)
028    Harry:   Well that la- g- guess what that last one is.
029    Ed:      Haha
030    Harry:   Which one haven't I said?
031    Paul:    Er (.) neighbour points.
032             ((General laughter.))
```

Although Paul's joke exploits a pattern ('learner speaks and teacher points' etc.), its impact depends on the familiarity of all concerned with the representation of classroom behaviours in the language classroom, particularly in pair work. So although the idea that a neighbour could take on the role of pointer is unlikely in the context of this particular explanation, what makes Paul's comment so effective in generating immediate and universal laughter is that it exploits a common format that all recognise. This is confirmed by Harry's comment when the laughter has subsided: 'No no neighbours in this. No.'

Deflecting concerns

Sometimes professional concerns are more pressing and less benign, in which case humour may provide a means not only of orienting to shared feelings but also of relieving tension. The following examples (4.16 and 4.17) from DOTS illustrate this dual function and have been chosen partly because of what they reveal about the positioning of the team leader vis-à-vis the group and the wider organisation of which it is a part. From an analytical perspective they also illustrate why it is important to examine fine details of the sequence within which laughter occurs.

A team meeting, which takes place two weeks later than the one analysed in Chapter 3, has progressed fairly smoothly when Julie raises a question about the new appraisal (PAD) system, asking whether any announcement has been made about how this will be linked to performance related pay. When John says that there will not be a link, Julie pursues the topic, broadening it out to the issue of pay awards generally:

Extract 4.16

```
001  Julie:  Or what any of our pay awards might be:=
002   John:  =No:=
003  Julie:  =based on-=
004  Penny:  =(x)hh!
005  Julie:  (xx ⌈xx bonus)
006    Sue:      ⌊°(What) ° pay aw ⌈a:rd.
007  Penny:                         ⌊HHHehhe°hehe ⌈heh°!   ⌉
008    Sue:                                      ⌊Gh!o:⌋:d.
009   John:  No THAt's >what I m-<=
010  Julie:  =There was one bit of unrealism:=ab(.)out
011         (.) PAhy award.=
012    All:  =((Explosive laughter))
```

The placement of laughter here is interesting. The initial exchange between Julie and John is straightforward, with John taking advantage of Julie's lengthened 'be:' at the end of a syntactic and semantic unit to interpolate a negative response into her query. The cut-off at the end of her utterance in line 3 is interactionally ambiguous, interpretable as a sudden termination of her talk occasioned by John's negative response, but nevertheless lacking the falling intonation that would signal turn completion. The placement of Penny's brief turn (l.4) treats it as complete and although Julie does carry on to add a reference to a bonus this turn is overlapped by Sue's question (l.6). Penny's contribution is hearable as incipient laughter, the catch at the beginning and the explosive outbreath being typical of a withheld laugh. This is certainly what Sue seems to be orienting to in her question, uttered with a laughing tone and falling intonation and therefore offered as a candidate laughable. As such it is successful, immediately prompting laughter from Penny. Up to this point, though, the laughter has been confined to only two participants and has been relatively tentative, involving only a suppressed explosion from Penny, no explicit laughter from Sue, and a rapid decrease in volume of Penny's laughter which indicates its early termination. Sue's exclamation in line 8 serves to confirm this and John's subsequent return to the original topic marks a switch back to a serious key. However, the introduction of laughter into otherwise serious business allows Julie to revisit her previous utterance with a re-evaluation ('one bit of unrealism') that prompts immediate and explosive laughter.

Even allowing for the fact that the topic of pay awards has been marked as the object of laughter, the sudden explosion that is latched onto Julie's humorous re-evaluation is unusual. Such outbursts, involving all participants, are very rare in the data. To some extent this is understandable: these are not prepared jokes with a definite punchline that is designed to elicit laughter immediately upon its completion (Norrick, 2003: 1345) but local alleviations of otherwise serious business along the same lines as Penny and Sue's initial exchange. Why then should this apparently innocuous statement prompt such an immediate and emphatic response? The fact that Sue and Penny had established a humorous context certainly helped, as did Julie's tiny signal of a candidate laughable in the slightly loud and breathy 'PAhy' after a micropause, but these are hardly unusual. To account for the strength of this response we need to consider this exchange in the light of current and ongoing concerns.

As the argument extract in Chapter 3 showed, pay arrangements are much on the mind of the team at DOTS. The parent company has not been doing as well as expected and rumours about closures and/or redundancies are in the air, so all changes are being watched with great interest. In this context, the company's recent introduction of a new appraisal system that nobody seems able to explain and its current equivocation on the subject of pay awards and bonuses are hot topics that have given rise to serious concern and frequent probing. These doubts and concerns were evident in the meeting that had taken place two weeks earlier (Chapter 3) and interest in them has grown rather than diminished. The general feeling is that bonuses and pay awards will disappear, though this has not been confirmed. In situations like this it is often Julie, effectively John's deputy, who speaks for the team and who puts pressure on John to provide them with the information they need. The tension created by John's negative response has not been relieved by Sue and Penny's brief challenge, but when John returns to it, beginning his turn (1.9) with a negative, Julie's interruption has particular force. Her comment picks up on Sue's earlier rhetorical question implying that there will be no pay award (1.6) and accepts that Julie's initial serious question might have included an unrealistic assumption. Innocuous as this may seem, her intervention as spokesperson for the group allows all those present to respond emphatically to her hyperbolic representation of a perfectly reasonable assumption by laughing out loud, thus upgrading the force of her implicit claim that it is the company, not those present, who are being unreasonable. Explained in terms of the functions of laughter, this would seem to be a clear case of relieving tension, but more than this is at issue: it is also a reaffirmation of the group's shared evaluation of the company's position.

As in the earlier argument, John finds himself in a position where he sympathises with the views of his team but nevertheless has to represent the position of the company as far as he understands it. His situation is not made any easier by the fact that he is almost as much in the dark as his colleagues, though as the next extract shows, he can exploit this uncertainty as a means of affiliating with the group. The exchange follows very closely on the heels of the previous extract, after a brief explanation by John that the PAD isn't connected to the annual pay round and in any case is only applied to those currently on pay progression:

Extract 4.17

```
001   John:   There is the question of:
002           (0.6)
003   John:   BOnuses,
004           (0.6)
005   John:   and if there were such a thing as a bonus,
006           (0.4)
007   Penny:  Tchh ⌈hhh!
008   Julie:       ⌊HHHhh ⌈hahahaha °h ⌉ ahh⌈ah°
009   Sue:                ⌊hhhhehhehh ⌋      ⌊
010   John:                                  ⌊then that
011           would PRobab ⌈ly  be  pad  li ⌉ nked.
012   Julie?:              ⌊°t!hhhhhahah°    ⌋
```

This time the laughter is not spontaneous, following the more usual trajectory of development and decline. John leaves relatively long pauses in his delivery, but not at transition relevance places. The first turn ends with a slightly extended 'of' which, given the lack of any indication of the subject, cannot be interpreted as an invitation for completion, and the other two turns have continuing intonational contours that serve to hold the floor. However, his representation of the bonus as hypothetical (l.5) echoes the ontological status Julia assigned to pay awards which prompted the earlier laughter. Again, it is Penny who first shows signs of laughter but this time it is picked up immediately by Julie and Sue, John completing his utterance only when there are signs that the laughter is coming to an end.

By representing the company's position in this way, John has aligned himself with the team in a shared view of the unreasonableness, or at least unreality, of the company's position. This represents an important affiliative move of the sort that can be achieved through the use of humour, where laughter in response to an intendedly non-serious utterance reveals 'a coincidence of thought, attitude, sense of humour, and the like' (Schenkein, 1972: 371). However, there are differences between the two examples. In the first the humour is effectively shared since the effectiveness of Julie's laughter-provoking reassessment depends in part on the earlier work of Penny and Sue, and the laughter that ensues is both spontaneous and universal. However, John's humorous qualification does not prompt an immediate response, the gap of almost half a second making it appear as though he is waiting for audience appreciation. And the laughter, when it comes, is brief and involves only three participants, with Julie contributing the lion's share. Humour here achieves its affiliative purpose, but the contrasting responses also serve to confirm John's slightly ambivalent standing in the context of the group as a whole.

Outside targets

As far as topics for humour are concerned nothing stands out more than the use of outside agents, whether individual or institutional, as the butt of jokes. Perhaps because direct discussion of work tends to be confined to formal meetings these represent the natural arena for such jokes, which are rarely to be found in more casual encounters except as part of back region talk. This will be discussed in the next chapter and the role of humour in the construction of external identities will feature in Chapter 7. The final part of this chapter will set the scene for subsequent chapters by demonstrating that, whatever the similarities in the workings of humour among these three settings when professional business is being discussed, less formal settings allow for the development of distinctive styles.

Things we find funny

A person's sense of humour is an aspect of their personality, perhaps intrinsic but certainly coloured by experience and outlook, sensitive to place and possibility. I can remember watching a film with people unsympathetic to it where my strained and stunted laughter died quickly and awkwardly in their silence, then seeing it again some months later with my brother and being reduced by the contagion of his laughter to utter helplessness. Humour, perhaps more than anything else, is influenced by the company it keeps, and Wenger's suggestion (1998) that different communities of practice develop distinctive ways of doing humour is immediately appealing.

So far this chapter has focused on similarities rather than differences in the ways the three groups handle humour, but this is to a large extent a reflection of the circumstances of its construction: meetings constrain what is possible and are not conducive to extended displays that allow full rein for distinctive characteristics to emerge. When the professional situation is left behind, though, such limitations disappear and the group is free to develop its distinctive style. This section will be based on a comparison of extracts from a single break in both DOTS and the Pen in order to bring out the distinctive differences between them and develop explanations that might account for these differences.

Each of the following extracts, presented together for the purposes of comparison, is taken from breaktime talk in a staff/common room and has been chosen because it is typical of the sort of humour that is to be found in this particular setting. The breaks involved are of roughly the same length (20 minutes in the case of the Pen and 30 minutes at DOTS) and although not all colleagues are present the participants are fairly representative of those who would normally be present during a break.

The Exhibits

Extract 4.18

This extract develops from Paul's comment that he's having a 'carrot and orange' party.

```
001   Paul:   =The juice is flowing.
002   Harry:  Yeah. (.) The organ juicies. hh hhh
003   Paul:   Organ jucies.
004           (3.0)
```

```
005   Paul:    hhhhh heheh
006            (1.5)
007   Paul:    °heheh° Couldn't believe that. Nearly wet
008            myself when I read that.
009   Harry:   HHH!
010            (1.0)
011   Susan:   What was that?
012   Paul:    This:: well this exam. I was marking an exam
013            you know >it was< (.) 'Pre- prepare a menu
014            (0.8)
015   Paul:    for your friends.
016            (1.0)
017   Paul:    °And they had you know a list (.) of things
018            (.) and then° drinks. 'Organ jucies.' °hhh ·
019            hh°
020   Harry:   hh hh hhh
021   Paul:    Orange juice.
022   Keith:   HOH HOH!=
023   Paul:    =Organ jucies.=
024   Keith:   =Organ jucies.=
025   Susan:   =Ni:ce.=
026   Keith:   =(Well xxxxx) Heh!
027            (1.0)
028   Keith:   A:::h=
029   Paul:    =And I was- (.) d-I was saying yesterday the
030            other one was er (.) em
031            (1.5)
032   Paul:    'Please don't come round because I'm having
033            a pussy flap installed.'
034   Ed:      Hehe ⌜hehehehehehh
035   Keith:         ⌞A pussy flap! Heh heh · hhh
036   Ed:      °(xxxxxxx)°
037   Keith:   Hah hah=
038   Paul:    =>What was that one< about the motorist, 'So
039            I stuck my fingers up him.
040   Harry:   Yes. ⌜It was.
041   Keith:        ⌞HEEEEH! HHHAH!=
042   Harry:   =Hehehehehahah
043   Paul:    'He was travelling too fast >so what was it
044            he was- he was on his bike.< So I stuck my
045            fingers up him.'
```

```
046   Keith:   Hheh!=
047   Susan:=  Hhahah!=
048    Paul:   ='I shall
049             ┌be ┐wearing a rose in my bottom hole.'
050   Keith:   └Heh┘
051   Keith:   Hhehhh!
052   Harry:   That's a good one, yeah.=
053   Susan:   =┌┌Heh what was that.           ┐
054   Harry:    [[ 'She will recognise me┘
055   Harry:   She ┌will recognise me┐ =
056    Paul:       └I will be wearing┘
057   Harry:   =easily because I will be wearing=
058   Keith:   =Heh=
059   Harry:   =wearing a red rose in my
060             bottom ┌ho:le.'
061    Paul:          └hh hh hh hh ┌hah HAH!
062   Keith                        └HAH HAH!
063    Paul:   Heh ha:::h.
064             (1.0)
065   Keith:   HHHHH HEHEHAHAH! Walking on all
066             fours ┌presumably.
067   Harry:         └Heheh hahahahahah
068   Susan:   hhhehe hahahah!
069    Paul:   I'll just stand ┌(here in the=
070   Susan:                   └Yeah I think=
071    Paul:   =middle of the room. )┐
072   Susan:   =we can imagine that  ┘thanks.
073   Harry:   Heheh
             ((34 lines omitted))
074   Harry:   We need a benchmark °(to)°
075             (5.0)
076    Paul:   W- what- what does the term 'benchmark'
077             come from?
078             (0.5)
079     Ed:    Surveying.
080    Paul:   Surveying?
081     Ed:    Mm
082             (1.0)
083    Paul:   Stand on a bench?
084   Harry:   Walk on a bench, yeah.
085             (1.5)
```

```
086 Annette:  A workbench ⌐is it?
087   Paul:              ⌐You mean- (.) like (.) 'How
088            far is it from here to that bench?'
089   Harry:  hh hhh=
090 Annette:  =Heheheh
091   Keith:  Heh heh
092            ((20 lines omitted))
093     Ed:   And it could have been e:r (.)
094            ⌐on a bench.
095   Paul:   ⌐Heheh that guy sitting on the bench.=
096     Ed:   =Something that doesn't
097 Annette:  Heheheheh yea:h heh
098            ((10 lines omitted))
099     Ed:   Could be (.) a clear line (xxxxxxxxx).
100            (3.0)
101   Paul:   Probably a bloke called Mark sitting on a
102            bench.
103   Harry:  (xxxxxxxxxxx)
104   Paul:   'Can you sit on that bench, Mark.'
105 Annette:  Hhhheh
106   Keith:  Hahah (.) ⌐hah
107   Harry:          ⌐hah
108 Annette:  hah
109   Paul:   hhh hh
110   Harry:  But don't move.=
111 Annette:  =A::::h (.) heheh
112            ((118 lines omitted.))
113   Harry:  ((Reading definition from a dictionary))
114            'A mark on a stone post or other permanent
115            feature,'
116            (0.6)
117 Annette:  Mmm
118   Harry:  'at a point whose exact elevation and
119            position is known.
120            (0.6)
121   Harry:  Used as a reference point in surveying.'
122   Paul:   From the Old English 'Can you sit on that
123            bench Mark please.'
124            ⌐°(xxxxxxxxxxxxxxxxx isn't it.)°
125   Keith:  |Hehah
126 Annette:  ⌐Hehehehehehehe
```

Extract 4.19

Tim is trying to get a biscuit out of the packet, but is
pulling out two instead of one.

```
001      Tim:   It's stu:ck (.) I can't get rid of it.
002      Sue:   You ⌈pi:::g!
003      ???:       ⌈ Yeh-!
004      Anne:      └ Hh!
005      Tim:   ((High pitched)) IT'S stuck to i::::::t!
006             (6.5)
007      Sue:   °I like those.°
008             (1.5)
009      Tim:   I don't like (two).
010      Sue:   But he eats 'em.
011             (0.6)
012     ????:   Hehehe>hehah<
013      Sue:   (Like my mum >she says<) 'Oh I don't like
014             these biscuits.'=But she eats 'em.
015             (2.5)
016      ???:   I hate bourbons.
017             (0.8)
018      Sue:   I lo:ve bourbons.
019             (2.0)
020      Joe:   Do you like custard creams?
021    Julie?:  mm ⌈mmmm
022      Sue:       └ MMMMMMmmmmmm (amazing)biscuits
023      Joe:   (I cn I ca-) o- o:ld people's biscuits,
024             °custard creams. (They're xxx)°=
025      Sue:   =You⌈calling us OLD? You ⌈ca⌉lled us ⌈fat.⌉=
026     Julie:     └(That's xxxxxx)⌋      └he⌋       └HEh!⌋
027      Sue:   =the other d⌈ay.
028     Julie:              └YouHH! Called us FAt the other
029             da::⌈y,
030      Sue:      └(Then you⌈ said (xxxxxxxx⌉ you little)
031      ???:              └(xxxxxx xxxxx)⌋
032     Julie:   And SAD!
033             ((Brief confused overlapping talk involving
034             two different conversations.))
035      Joe:   HEH!HEhhehhehOH! I shan't-
036             (0.4)
```

```
037   Joe:   say anything.
038          (0.8)
039   Sue:   PLEa:::se!
040          ((The conversation continues sporadically,
041          then about ten minutes later, during a
042          period of silence, Tim exposes his stomach.))
043   Sue:   OcH ⌈H!
044   ????:      ⌊(°xx ⌈xx°)
045   Sue:           ⌊EVe ⌈ry
046   Tim:               ⌊NO::!< I
047          >just< ⌈wanted to s⌈ee ⌉ how fl⌉at my=
048   Sue:          ⌊blooming ti│ me.⌋        │
049   ???:                      ⌊Honestly! ⌋
050   Tim:   =stomach w⌈as.
051   ???:             ⌊Every si⌈ngle time.⌉
052   Sue:                     ⌊Every (xxxˡxx) ti::me.
053   Tim:   ⌈⌈(xx) ⌈I (xxxxxXx)⌉
054   Sue:   ⌊⌊°I tell you what°⌋ I'm going to start getting my
055          stomach out⌈soon. (xxxxxxxxxxxxxxx)⌉ shy::.⌉
056   ???:              ⌊Yeah we'll ⌈all start-      │      │
057   Tim:                          ⌊C(x)HH⌋HAh! ⌋ hahhah⌋=
058   Tim:   =hahah ⌈ahahhahah ⌈hhahhahh       ⌉
059   ????:         ⌊hehehahha │ hah     ⌉     │
060   Sue:                     ⌊I'll⌋ get⌋ my sto ⌈mach
061   Tim:                                       ⌊They've
062          got that on tape Sue.=HU:H (.) huhhuhhahhah!
063          ((General laughter and chuckling.))
```

Different style . . .

The differences between these two are immediately apparent and quite striking, the impersonal, professionally oriented exchanges of the first contrasting markedly with the freely traded insults in the second. In fact, the differences go deeper than this, permeating the interactional and interpersonal construction of the humour, as even a brief summary reveals.

The first part of the Pen extract (4.18) develops in three phases separated by evaluations. The first episode arises from Harry's mention of an amusing linguistic error, 'organ juices', in line 2, which allows him to share laughter with Paul, further extended when Susan invites an explanation. Her evaluation of this as 'nice' in line 25 marks a shift to a new phase in which Paul and Harry quote more examples of amusing expressions beginning with the merely suggestive 'pussy flap' and moving to the more explicit 'stuck my finger up him' and 'wearing a rose in my bottom hole'. When Harry evaluates this as 'a good one' (l.52) Susan again invites an explanation, and this time Harry's response is followed by extensions from both Keith and Paul. After a short exchange relating to student placement, the second part begins with a question about the origin of 'benchmark'. The answer to this develops over considerable time and sees the extension of the joke from the initial playing with the word 'bench', through the idea of a man sitting on a bench, to the joining of 'bench' and 'mark' to create a humorous etymology, which is finally contextualised (considerably later) as the completion of a genuine definition. The initial awkward piggybacking of the spurious source on the genuine one is finally realised in a linguistically seamless join which serves to close the topic on a humorous note.

The relative elegance of this is in stark contrast to the earthier exchanges in the DOTS common room, where the humour depends on personal baiting. The first example here is also divided into three phases beginning with an initial insult from Sue directed at Tim who has accidentally taken two biscuits. After period of silence, the evaluation of different biscuits produces a further dig at Tim (l.10), who ceases to become the butt when Joe inadvertently criticises Sue and Julie who have expressed their liking for custard creams (ll.23–4). From this point onwards, Joe's previous insults are paraded before him in mock indignation by the pair he has unintentionally 'offended'. Mock offence is also the defining characteristic of the second part, where Tim's exposed stomach elicits first pretended disgust and then a threat to copy his action.

Topic

This is the most obvious difference between the two extracts and it may not be entirely incidental. Interest in both parts of the Pen extract falls on language, albeit in different ways, the first part effectively comprising a list of entertaining linguistic mistakes and the second an extended exploration of possible etymologies which provides the opportunity for amusing proposals. As language teachers the participants have an interest in language and access to a stock of relevant examples. This contrasts markedly with the physical and behavioural orientation of the second extract, with its focus on greed and bodily features. The physicality of the humour is typical of interaction in the DOTS but alien to the Pen culture. It is also a topic that participants are happy to exploit at their own expense, as the following extract from a different break demonstrates. The head of the Unit, John, has just asked Margaret if she has any electronic pictures of her recent visit to Nepal. At first she says she has none then suddenly remembers that she has just one picture:

Extract 4.20

```
001  Margaret:  Well no:=that's not quite true I've got
002             one picture
003             (0.5)
004  Margaret:  of Nepal,
005             (0.4)
006  Margaret:  that's electronic that was of a (.)
007             farmers' meeting.
008     John:   Oh that's okay can I have a co ⌈py
009  Margaret:                                 ⌊No you
010             can't.= ((Uttered very decisively and
011             abruptly.))
012     ???:    =GHh ⌈heh!
013  Margaret:       ⌊Because I've got a big bum in it.
014             ((General extended laughter.))
```

Style

Closely allied to the topic is the style of the humour, again markedly different. Humour in the DOTS extract depends on teasing, or more specifically insult and reaction to insult. Within 32 lines, the terms *pig*, *fat*, *old* and *sad* have all been tossed into the arena and the first part ends with a plea for Joe to shut up. By contrast, the humour in the Pen tends towards linguistic wit, either crudely presented as in the list of examples, or more subtly proposed as with Paul and Harry's allusion to making a mark on a bench (ll.83–4), Paul's introduction of 'Mark' to provide a cod definition (ll.101–4) and his false etymology (l.122).

Delivery

While the Pen delivery is detached, as befits the tradition of wit, exaggeration is the hallmark of DOTS humour in this extract. This is not merely a matter of expressive delivery in descriptions like 'pi:::g!' or 'SAD!', but the cumulative impact of repetition exemplified by the hyperbolic 'every blooming time', 'every single time' and 'Every (xxxxx) ti::me' in the second part, the force increased by what could be described as tandem delivery.

Direction

While colleagues are the targets for humour in DOTS, in the Pen the humour is either neutral or directed at outsiders, in this case students and exam candidates who make linguistic errors.

Orientation

In directing their humour at colleagues, participants in DOTS set up conflictual situations. The status of personal insults as a routine matter in the DOTS common room undermines their face-threatening potential, but the dominant mood is nevertheless oppositional. The second part derives entirely from Sue's opposition to Tim's bare stomach, but the first part also provides ample evidence of this orientation in the oppos- itional pairs 'like/don't like' (ll.7–9), 'hate/love' (ll.16–18) and 'amazing biscuits/old people's biscuits' (ll.22–3). By contrast, the Pen exchanges are marked by co-operative moves such as supportive repetition (ll.2–3, 23–4, 83–4), invitations to expand (ll.11 and 53) and positive evaluation (ll.25 and 52).

Organisation

The oppositional orientation characteristic of DOTS exchanges necessit- ates a pursuit of points through persistence in the face of overlapping talk or contrasting positions, as in the competition for the floor between lines 45 and 50 of Extract 4.19:

```
045  Sue:              EVe ry
046  Tim:                 [NO::!I
047       >just< wanted to s ee   how fl at my=
048  Sue:           [blooming ti me.]
049  ???:                         [Honestly!
050  Tim:  =stomach was.
```

By contrast, the extended topics in the Pen extracts allow space for humorous comments that reprise and build on earlier contributions, a form of shared construction that is typical of Pen talk.

... different place

In order to account for the way humour works in these two settings and to account for the differences in style, it will be necessary to consider it in the context of other forms of talk that will be the business of the next two chapters. However, as a preliminary to this it is possible to identify specific factors that may contribute to the differences.

The first and most obvious is the setting. The Pen staffroom is a place where business and pleasure combine; it is where teachers do all of the work (administration, preparation, etc.) that is not covered in the classroom and it cannot therefore conveniently be separated from their professional world. While an element of humorous conflict might be acceptable, the working context means that this could all too easily tear against the grain of the professional imperative, jeopardising the collaborative orientation that accounts for so much of its success. In this sense, it is at least possible that some of the restrictions associated with humour in meetings apply here. This is an aspect of back region behaviour that has not been considered and one which suggests that the contradiction of front region activity may bring with it certain constraints associated with the professionalism that makes it possible. Actors may set aside their professional fronts but what lies behind these must maintain its integrity.

None of this applies to the DOTS common room, which is a place of retreat from work, a location where professional concerns can be laid aside and entirely personal topics pursued. Colleagues can therefore indulge in insults and banter without jeopardising professional decisions, and the greater the distance they are able to establish between workplace and retreat the more effectively the latter fulfils its primary purpose and the better they are able to build personal relationships within it.

The absence of a common room in IEC accounts for its omission from this final section, there being no area set aside for the group to meet in breaks or at other times. The privilege of individual offices not enjoyed by colleagues on either of the other two sites allows space for recreational talk, but this is essentially private and the group itself must take whatever opportunities occur prior to meetings or in the corridor.

Despite this limitation off-the record exchanges do take place in the IEC office, which is also where tea and coffee are made and where conversations on non-work topics sometimes develop. Perhaps because of the transitory nature of such encounters, humorous exchanges here often take the form of banter similar in some ways to the teasing that takes place in the DOTS common room. In a meeting exchange examined earlier the director, Jack, was teased by colleagues for an earlier mistake in timing and here the tables are turned as he allies with Kathy to tease Sally, his secretary. He begins by asking about an examination these two colleagues had supervised the previous Friday but is quick to take up the opportunity offered by Kathy's comment on Sally in line 15:

Extract 4.21

```
001   Jack:   Did it go o↓KAy on Fri̲day.
002           (0.8)
003   Sally:  Yeah (.) they were
004           ve̲ry GO̲ ⌈od.
005   Kathy:              ⌊Oh yeah they were ⌈lovely⌉
006   Sally:                                 ⌊It was ⌋lovely=
007   Kathy:  =Mmm ⌈mm
008   Sally:       ⌊yeah.
009           (0.4)
010   Sally:  They were ve̲ry good.
011           (1.0)
012   Kathy:  Yeah (.) it wa̲s good.
013   Jack:   Grea:t.
014           (1.5)
015   Kathy:  She beha̲ved herself a̲nyway Jack.
016           (1.0)
017   Jack:   Go̲od! It's more than she's do̲ne
018           this ⌈ mo  ⌉ rning by the sound ⌈of i⌉ t.
019   Kathy:       ⌊ She-⌋                   ⌊No.⌋
020   Kathy:  She's fee ⌈ling (xxxxxxx xxxx)⌉ it (down).
021   Sally:            ⌊°(xxxxx) hheheh-!°  ⌋
022   Jack:   Sheh's HHheh!
023           (0.2)
024   Sally:  °She's° beLLIGerent!=this morn ⌈ing.
025   Jack:                                  ⌊She's
026           GOT- a co̲b ⌈ on.
027   Sally:             ⌊ HOrmones:.=
028   Jack:   =She's go(x)a co̲b on.
029   Kathy:  That's ⌈it.
030   Sally:         ⌊It's ho̲ ⌈rmones.
031   Jack:                    ⌊It̲'s ↓ho̲:rmones.
032           (0.2)
033   Jack:   Is̲ it?
034   Sally:  Yeah-.
```

```
035    Jack:   I wish I could have some o' these
036            BLOODY hormones (.) give me an excuse
037            to ha ┌ve a good┐ rant.
038    Sally:       └Ho:rmones┘
039            (0.8)
040    Jack:   Every time I get a good- >have a good rant
041            it's because< I'm a ↓ man.
042    Sally:  Ye ┌ah.
043    Kathy:     └Hehh!Hha: ┌:h!
044    Jack:                 └If you're a woman you've
045            got 'o:rmones, if you're a man it's-
046            (0.4)
047    Jack:   cuz y're=a (.) man.
048    Kathy:  Heh!=hehhehe ┌hehheh!
049    Jack:                └Gerr: ┌:::┐ bli: ┌mey.
050    Sally:                       └ We-┘      └ I'm glad
051            you've-
052            (0.4)
053    Sally:  CO ┌ttoned on┐
054    Kathy:     └(xxxxxxx)┘ hormo ┌nes.
055    Jack:                        └HEH!ha:hh!=
056    Sally:  =I'm ghhlad he's aWARE of the
057            situahhtion.
058    Kathy:  HehHEHheh-!
```

This sequence falls into four quite distinct phases:

1.(1–13)   A series of assessments related to the students and the exam-
           ination.
2.(15–28) A series of teasing assessments of Kathy, all but one begin-
           ning with 'She', including Kathy's self-reference (l.20).
3.(35–47) Following Kathy's confirmation of her state (l.29) and Sally's
           proposal of a hormonal explanation (for a discussion of this
           as a topic for humour, see Wennerstrom, 2000), Jack's mock
           complaint about the unavailability of this as an excuse to
           him, which produces laughter on the part of Kathy.
4.(50–8)   Sally's repeated assertion of satisfaction with Jack's new
           awareness, ending in shared laughter at his expense.

Bearing in mind the potentially divisive effects of teasing, it is interesting
to study the shifting alignments here. In the first phase it is Sally and

Kathy who align in their judgement of the examination session as 'very good', then in the second phase, initiated by Sally, she aligns with Jack in a teasing judgement of Kathy's state. However, when in the third phase Jack selects the membership category 'male' as the basis for his humour, he thereby separates himself from the category to which the others belong, thus setting up the final phase where Kathy and Sally align at his expense. In a relatively short space the power of teasing to isolate an individual has been diminished by these shifting alignments and, with them, changing targets.

In complaining about his condition as a male, Jack is also mining a standard vein of humour in the relationship between him and his two female colleagues in the office, one they in their turn will exploit slightly later when the topic has shifted to recruitment, which has taken a worrying downturn. Jack points out that things may not be as bad as they seem and both he and Kathy have just agreed that 'we'll get there':

Extract 4.22

```
001   Jack:   I'M not the worrying sort.=
002   Kathy:  =Well we'll send you on the street if you
003           don't.=
004   Jack:   =That's right.
                      [
005   Sally:         We won't get much will we:?
006   Kathy:  Heh!Heh>hehe heh<
                           [
007   Jack:               Yehwhah!? Ihhf hhhI-
                                   [          ]
008   Kathy:                        Heh!heh!
009          (0.4)
010   Jack:   I 'ea::rd that.
```

Relationships such as this raise the delicate issue of how far individuals are responsible for the patterns of humour to be found in groups, and an examination of humour in IEC and the Pen suggests that they may have a significant influence. Sue, for example, is a dominant figure in both parts of the DOTS extract and her teasing humour often initiates exchanges of the sort seen in Extract 4.19. Although the evidence of that extract demonstrates that physical characteristics are fair game, her exaggerated and deliberatively provocative style encourages a style of teasing that is certainly more animated than the norm.

The same might be said of Paul in the Pen, who also happens to be a writer of jokes featured on national radio. His apparently conscious efforts to create humorous opportunities mean that his joking is not simply a part of the normal humorous interplay in the staffroom, and his natural, even professional, inclination to test his wit explains why, unlike any of his colleagues, he is happy to pursue a joke. In terms of topics, style and orientation his approach seems no different from that of colleagues, but the ongoing construction of humour and its energetic pursuit may owe much to his – and in the case of DOTS, Sue's – particular predilections.

The role of the 'joker' has been recognised in workplace humour (see, for example, the part 'Joshua' plays in the joking sequences described in detail by Handelman and Kapferer, 1972), but the relevance of this to the development of a distinctive style has not been explored. Without wishing to make too strong a claim for the importance of this, I think it should nevertheless be borne in mind when analysing workplace humour and drawing general conclusions about dominant patterns; otherwise, the joker in the pack may turn out to be just that.

A less contentious influence is the professional context and here the Pen teachers' interest in language is all too evident. It serves as both a topic for discussion and a natural source of humour, a professional source to be mined as the mood takes them. As agricultural scientists DOTS colleagues are perhaps less linguistically sensitive, but by the same token they are less physically sensitive. Their world brings them into intimate contact with the digestion of animals and all the indiscriminate filth of the countryside, they change from white coats to Wellington boots without giving this a second thought, and they are happy to muck in together to overcome the challenges and crises associated with an unpredictable environment. If their humour then turns on boisterous interplay around physical and behavioural characteristics, this is hardly surprising.

The picture that emerges, then, of different ways of doing humour is a complex one. Doubtless Holmes and Marra are right when they claim (2002: 1688) that '[d]istinctive styles of workplace humour develop, permitting comparison between different workplaces on this dimension', but comparisons based on typologies might easily miss the subtleties of design and presentation that contribute to the distinctive stripe of each setting. The claim that cohesive and non-cohesive groups have different humour patterns (Duncan, 1984: 895) may also be misleading. This chapter has shown that different cohesive groups have different patterns both between groups and within groups, depending on the particular setting and its relationship to doing professional business. The interplay of different factors in establishing a humorous environment is complex and resists definitive characterisation, though this is also an essential aspect of humour itself: to understand it fully is to miss the joke.

Conclusion

Humour, as Linstead (1988: 123) observes, 'is an essential and important part of organisational life', and it will reappear in various guises in the chapters that follow. The evidence of this chapter suggests that, where collaborative groups are concerned, it tends to take very similar forms in professional encounters, but outside the professional arena significant differences may emerge.

This perspective qualifies general claims about similar patterns in cohesive groups by drawing a distinction between formal and informal encounters, but to understand better the nature of the relationship between these two it will be necessary to explore more carefully features of a world where professional and personal meet: the back region. In doing so traditional perspectives on humour may also need to be recon-figured. These have adopted different approaches and utilised different typologies but a common distinction is that between control and resist-ance (e.g. Powell, 1988), also characterised as control/conflict (Steph-enson, 1951) and control/social tension (Paton, 1992). As the next chapter shows, there is another very important dimension that is not adequately represented by either side of this characterisation.

5
The Interactional Dynamic: Stories from the Back Region

Introduction

The stories lying at the heart of the workplace culture of the Pen teachers represent a confluence of experience and belief and offer insights into their engagement with the mysteries of their profession. In this chapter I hope to show, through an exploration of these teachers' tales, how the mystery of *their* work sets them apart from other groups, and how their modes of sharing represent both a commitment to the seriousness of their profession and a defence against the helplessness it can engender.

This decision to dedicate a chapter to only one of the three groups represents an attempt to capture some flavour of its essential nature, an ambition perhaps as delusional as the alchemist's quest for the quintessence. Nevertheless, what makes the effort worthwhile is the way in which the Pen staffroom functions as a back region, an interactional alembic that yields valuable glimpses of the relationship between professional ways of doing and being. The chapter will reveal how talk in this physical setting is constituted in very particular ways, marking it out as a place where humour is the default key and where sanctioned topics permit interactional behaviour that is conventionally proscribed. Here, more than anywhere else in the data set, is compelling evidence that long-established groups can unknowingly create unique conditions within which to explore their professional world.

An understanding of why this should apply only to the Pen group and not to the other teams in the study depends on the distinction drawn in the opening chapter between *back region* and *backstage*. It was suggested there that the former would be used, following Goffman (1959/1971: 114), to represent 'a place, relative to a given performance, where the impression fostered by the performance is knowingly contradicted as a matter of course'. This will take place backstage, a territory much more extensive and much more varied, covering any place in the institution not designated for professional–client encounters. Although Goffman tends to treat these two as synonymous, there are important differences between them which account for the special status of Pen staffroom interaction.

All the interaction analysed in this book takes place backstage, away from clients, but only the Pen staffroom represents a back region, as a consideration of the three different professional configurations makes clear. IEC is an academic unit specialising in distance learning courses, although some of its members are involved in face-to-face teaching. For most of the group teaching encounters take place overseas on study weekends or for a fortnight in-house on the summer school; for the rest of their time, business is mainly conducted through meetings, email or occasionally telephone encounters, and solitary writing. For this group professional–client encounters, geographically and technologic-ally distributed, are realised in collections of individual relationships that may range in terms of intensity from those that consist entirely of taped feedback on an assignment, through a classroom encounter over a single weekend, to relationships developed through email and face-to-face encounters, formal and informal, over five years. 'Performances' located in single, identifiable spaces are therefore probably more the exception than the norm. Even more significantly, there is no common room in which to meet, so there is quite literally no physical basis for a back region. It may be true that elements of back region behaviour emerge in meetings in individual offices, before and after meetings and in corridors, but this is only to be expected and does not define the character of such encounters.

DOTS colleagues do occasionally meet clients in their professional setting, but this is not the norm; encounters in the field, at conferences or on visits are more typical. There might even be a case here for arguing, with Sarangi and Roberts (1999: 23), that the laboratory and farm are the frontstage of professional practice, but this seems to be an unne-cessary complication: it is surely more straightforward to recognise that colleagues do most of their work backstage. Unlike the IEC group, they have access to a common room which is widely used during morning and afternoon breaks as well as during the lunch period. However, there is no evidence here of any back region behaviour, perhaps because clients are so far removed from this setting. There is no bar to discussing work in the common room, but judging by the infrequency with which it occurs as a topic, people here are only too happy to leave it behind.

Since job differentiation is much greater here than in the other two settings, colleagues might have less in common, or it may simply be that they are glad to put behind them the painstaking and sometimes repetitive work of the lab and animal shed.

The Pen is very different. Here, the staffroom is just across the way from classrooms and teachers escape to its sanctuary with the sweat of lessons still fresh on their brow. It is small wonder that Goffman himself chose the school staffroom as an example of a back region. The Pen staffroom, perhaps because of its mixed composition, does not manifest any of the grosser physical characteristics of a back region, but the teachers' talk nevertheless fits in well with Goffman's description. For example, students, typically the butt of humour, are represented in ways that might give serious offence if heard by, or passed on to, those targeted. More interesting than the topics and targets of talk, though, is the deployment of humour and stories as a professional resource that depends on the freedoms available in the back region – and the ways in which this is interactionally constructed to produce a distinctive cast to this unique professional space.

Using humour

By way of a general background I offer a rough picture of staffroom humour in the Pen, drawing on Paton's categorisation of humour in the workplace (1992: 5–6). Although there are numerous problems with any attempt to categorise humour, most typologies are very similar and the aim of Table 5.1 is to give a sense of the most common forms of humour in the staffroom, listing all those where I was able to identify at least five instances overall. The full list includes aspects such as fooling, derogatory nicknames, kidding, parodying, ridicule, sarcasm and insults, but occurrences of these are very rare.

Table 5.1 Main types of humour in the Pen staffroom

Type	Break	Meeting	All
Anecdotage, funny stories	27	4	31
Teasing	18	9	27
Parodying	4	1	5
Raillery, disparagement	3	4	7
Witticisms, banter, repartee	43	36	79
Total of the above	95	54	149
Total (all types)	**110**	**62**	**172**

As far as the Pen staff are concerned, humour is an important aspect of their professional life. As an illustration of this, consider what they look for in a colleague:

> The ones that make it here are the ones that don't take themselves too seriously.
>
> (Jenny)

> It might be something to do with how seriously you take yourself, actually ... The people that do fit in are not necessarily the people that you would expect to.
>
> (Harry)

> I think we do, we send out signals to people – and in a sense I do the same thing as I do in a class. I may make a cheeky comment which is obviously not meant to be- which is 'Eh look, this is only a joke.' I think we try to relax people, we definitely do.
>
> (Paul)

Additional emphasis, if it is needed, is provided by this comment about a teacher who did not fit in:

> He always looked so bloody miserable. That was the problem with him.
>
> (Paul)

A further example, this time from staffroom talk, demonstrates that the Pen staff are not only aware of the part which humour plays in their professional life, but also conscious of the form it often takes:

Extract 5.1

```
001     Harry:  The amount of deliberate misunderstanding
002             that goes on in this staffroom=
003  Annette:  =Heheh=
004     Keith:  =HEHEH HEHEH!
005     Harry:  Heheh
006      Paul:  What do you mean by that.
```

The most striking contrast between this and the other two sites lies in the almost complete absence of aggressive humour in the Pen staffroom. Compared with the insults, teasing, etc. that characterise the other sites, the exchanges here emerge as remarkably benign, amounting to only eight instances of aggressive humour, five of which involved ridicule. There was only one example of obscenity (in Extract 5.2 below), one of sarcasm, when Harry describes a list of language exam scores as 'exciting', and one insult, when Paul describes a group of students as 'three-year olds', then 'four-year olds' acting as five-year olds.

This mild orientation does not limit the range of humour, which takes many forms and functions in a variety of ways, but by far the most popular source of amusement and anecdote is the behaviour of students. It is here that the professional engagement of the staff is most directly realised, and it provides a valuable insight into the part that humour plays in their work as it is invoked in staffroom talk.

If relations with students touch the very heart of teachers' professional identity, the findings of research in this area do little to bolster the prestige of the profession. Hammersley suggests that in the 'semi-fictionalised world' of the staffroom he studied (1980: 54) features of students are amusingly accentuated in a way that involves unpleasant caricaturing, perhaps because 'a recurrent topic for comment and discussion in the staffroom is the "failure" of pupils to conform to those [teacher] expectations' (1980: 211). More than 10 years later Kainan did little to revise this view, insisting that the 'villain of these [staffroom] stories is always the student or group of students' (1992: 446).

Things in the Pen are rather different, the gentle tone of the humour extending also to relations with students There is only one example in the data of a genuinely angry outburst, which also includes swearing and potentially offensive national stereotyping:

Extract 5.2

```
001   Paul:  Can we have a notice on the door, (.) Harry. I
002          mean (.) it really pisses me off that
003          (0.5)
004   Paul:  this is typical (.) it's a typical Japanese
005          thing, (I mean it's just incredible (.) and
006          everybody needs to)
007    Ed:   Yeah.
008  Susan:  Mmm
009  Harry:  Yeah.
010          (1.0)
011    Ed:   Once the ball starts rolling
012          (2.0)
013  Keith:  (xxxxx) the biggest
```

```
014           signs ┌that you can (xxxxxx).
015    Paul:        └'No you cannot switch.'
016   Harry: Yeah.
017      Ed: Yeah.
018   Keith: (xxxxxxx) somewhere else.=
019    Paul: ='No you bloody well can't.'
020           ((General laughter.))
```

Humour functions here to mitigate the force of Paul's initial comment, which arises from his frustration with students who constantly press teachers to allow them to change classes, even though their initial placement is constantly reviewed. In this case the switch from genuine to mock anger produces the necessary resolution. Paul's first complaint is addressed to Harry in the latter's role as academic manager during Jenny's absence, and his outburst receives an immediate and sympathetic response from all the Pen teachers present. Following a short pause Ed then offers an implicit justification for Paul's response by suggesting that if it is not stopped now the problem will escalate (l.11). Paul's return to the issue (l.15) picks up on his earlier call for a notice on the staffroom door and responds to Keith's reference to this by providing a text, which is then recast in humorous form by reprising his original anger as mock-anger. The emphasis and the deliberate inclusion of (milder) swearing achieve the necessary effect: this creative redefinition still expresses disapproval of the student's action but in a way that relieves the tension created by Paul's original formulation of this disapproval.

More typically, where students feature as the butt of humour it is often in the much milder guise of disparagement which seems designed more for comic effect than as a reflection of dissatisfaction or irritation, as in Paul's comment in Extract 5.3:

Extract 5.3

```
001  Annette: Well I managed to get it across with my
002           group, which I think is fair ┌enough
003    Harry:                              └Yeah
004    Harry: Yeah
005     Paul: Because if anything's going to throw them
006           it's (.) an instruction.
007    Harry: Heheheheh. hhhh hhh
```

As the last chapter showed, humour features in all three backstage settings and the only thing remarkable about Pen humour is its relatively innocuous character. In a curiously contrary way, what is more interesting in this staffroom is what it takes to be serious.

Being serious

There is plenty of evidence in the data to show that humour establishes different perspectives on professional issues and contributes to the social dynamics of the staffroom through its power to establish distance or generate involvement, but this does not in itself provide evidence for its centrality in this particular collaborative culture. However, if it can be shown that it is the absence of humour rather than its presence that requires explicit signalling in this back region, this will provide persuasive evidence of its significance. In what follows I argue that humour is indeed the dominant key in this professional environment and that any shift out of this key needs to be marked by a specific contextualisation cue: the use of direct address.

Hymes proposed that a particular key might be conventionally associated with a particular setting, person etc., and that there might also be conventional ways of marking a change of key:

> Key is often conventionally ascribed to an instance of some other component as its attribute; seriousness, for example, may be the expected concomitant of a scene, participant, act code, or genre (say, a church, a judge, a vow, use of Latin, obsequies). Yet there is always the possibility that there is a conventionally understood way of substituting an alternative key.
>
> (Hymes, 1986: 62)

The *contextualisation cue* identified by Gumperz is a broader linguistic marker, which may in some cases serve to bring about the shift to which Hymes refers. Gumperz defines this as 'any feature of linguistic form that contributes to the signalling of contextual presuppositions,' having described such cues as 'the means by which speakers signal and listeners interpret what the activity is, how semantic content is to be understood and how each sentence relates to what precedes or follows' (1982: 131). He also points out that '[f]or the most part they are habitually used and perceived but rarely consciously noted and almost never talked about directly' (*loc. cit.*). This describes very accurately the use of direct address to signal serious business in the Pen staffroom.

The immediate evidence in support of this emerges from an analysis of all instances where direct address is used. There are, in fact, surprisingly few examples of this and only three occasions in which humour features, all easily accounted for. The first occurs in the first line of Extract 5.2, but at this point Paul is angry and serious; it is only later, following the introduction of the idea of a sign on the door, that the key changes. In the other two examples, the humour actually depends on the speaker playing off the use of direct address as a marker of seriousness and thus reaffirming its status:

Extract 5.4

```
001  Annette:  Harry, I think a couple of little gr ai::ns
002            of sugar might have fallen in your (.) as I
003            took the spoon over for my coffee.
004    Harry:  Did it?
005  Annette:  °Hehh°
006    Harry:  We::ll I'll try it °but er° (.) and get back
007            to you.
008  Annette:  Hehehe ┌heheheh
009    Harry:         └Heheheh
010            (4.0)
011    Harry:  That will be all right thank you.=
012  Annette:  =Heheheheh. Heheheh
```

Annette has spilt 'a couple of little grains of sugar' in Harry's coffee, but the 'serious' business of indirect apology is undermined by the bathetic nature of the offence, signalled by the exaggerated delivery of 'grains'. Harry orients to this first by using a formulation appropriate to important matters of business but incongruous in this setting, then by providing an explicit evaluation. The humour here depends on treating a trivial accident as a matter calling for proper procedure and a formal response. The other example depends on a more explicit parody of a serious request, with Susan exaggerating her helplessness in Harry's absence.

Conclusive evidence that direct address is designed to signal serious business is to be found in the two instances in the data where an unsuccessful attempt is made to introduce a humorous note following the use of this cue. The first involves an interruptive pun:

Extract 5.5

```
001  Harry:  Paul- (.) Paul and Ed. hh hhh
002          Well ┌Ed al ┐ ready knows.
003   Paul:       └Yeah. ┘
004   Paul:  Yeah.
```

```
005    Harry:   (All that came out) with Nina. E:m
006             (1.0)
007    Harry:   she was saying that the- the two Poles
008     Paul:   Yeah.
009    Harry:   particularly:
010  Annette:   °North and South.°=
011    Harry:   =Grace,
012     Paul:   Mm
013    Harry:   er tend to (.) jump in
```

Puns, as Norrick notes (2003: 1345–6), 'count as frivolous and superficial even among the various types of humour, and they rank quite high on the scale of aggression, because they disrupt topical talk by misconstruing and redirecting it' and Annette here is an outsider not included in Harry's direct address. Even so, Harry's response to her comment is brutally dismissive. Annette's pun is delivered quietly, but close attention to the turn within its interactional sequence leaves no doubt that it has been deliberately rejected. The lengthened vowel at the end of 'particularly' (l.9) provides Annette with an opportunity to seize the turn at what is not a transition relevant place (Sacks et al., 1974), but she is not interrupted. Instead, Harry leaves her to complete her turn and latches his continuation onto this (l.11), thus allowing space for her turn but then ignoring it. Annette's attempt at humorous interpolation has been downgraded to the status of a mere continuer.

The second example is notable for the way in which the attempt to introduce a humorous anecdote is pursued in the face of clear signals that it is inappropriate. After a short pause following an extended humorous exchange between Paul and Susan that had involved much dramatisation, Annette switches rather abruptly to serious business by addressing Susan by name and inviting her and Khaled, her student, to join Annette in the listening centre. Susan agrees then introduces a story preface (parts of the exchanges are missing because of a parallel conversation taking place close by):

Extract 5.6

```
001    Susan:   Well-┌it might be a use┐ful thing to=
002  Annette:        └Of cou::rse, yes.┘
003    Susan:   =do in his (xxxxxxxxx) because yesterday
004             Harry (xxxxxxxxxxxxxxxxxxxxxxxxxxxxx
005             with it really.)
```

```
006  Annette:  Yes well
007            (0.5)
008  Susan:    >I said< 'Well did you listen to the tape?'
009            ((Imitating Khaled's accent)) 'Oh no I no
010            listen to tape.'
011  Annette:  Oh right.=
012  Susan:    ='Okay, what did you do?' 'I read
013            something.'
014            (0.5)
015  Susan:    'Yeah. Alright.' HHHH!
016  Annette:  (xxxxxxxxxxxxxxxxxxxxxxxxxxxxxxx)
017  Susan:    (xxxxxxxxxxxxxxx) Hehehah
018  Annette:  (xxxxxxxxxxxxxxxxxxx)
019  Susan:    He usually just takes a quick look at the
020            words and ┌(xxxxxxxxxxxxx)
021  Annette:            └ Mmm I see.
022  Susan:    Yeah. So I don't quite know what ┌to do┐
023  Annette:                                   └Okay ┘ so
024            that might be quite a good
025            idea ┌then.
026  Susan:         └ Yeah that'd be great. =
027  Annette:  =So if we go down to the listening ce:ntre
028            (.) with the students, and I'll show you
029            and (Jenny) how it *works*.
030  Susan:    *Yeah.*
031            (2.5)
032  Annette:  >Thank you very much.<
```

Susan uses the opportunity of evaluating this visit (l.1) to announce an anecdote via the use of a causal conjunction ('because' in line 3). She ignores Annette's clear attempt to close this down and shift back to serious business ('yes well'), going on to dramatise an exchange with Khaled. The imitation of Khaled's accent succeeds only in eliciting 'Oh' from Annette (l.11), a change of state token (Heritage, 1984a) indicating the receipt of new information. Since Susan's dramatisation is designed not to provide information but as form of entertainment, the pragmatic force of this deliberate misreading is clear. At this point Susan can bring the anecdote to an end on a suitably non-humorous note, an opportunity that occurs again in line 15 following the lack of any response to her dramatised exchange. Instead she offers an assessment of Khaled's response, the laughter tokens of a loud exhalation of breath at the end of this turn and her subsequent laughter (l.17) indicating that the imitation of Khaled's accent is offered as a candidate laughable. When this fails to generate laughter, Susan produces a quick, non-humorous summary of the problem (l.19), which does count as new information and is responded to appropriately by Annette. Finally, Susan offers an *upshot* ('So I don't quite know what to do.'), allowing Annette to shift back into the original topic (ll.23–5) and then make arrangements explicit. Annette's brisk closure (l.32) serves to further emphasise that this has been a business exchange.

These exchanges are significant not just because they provide compelling evidence that there are situations in the staffroom where humour is inappropriate, but because the signalling that marks it is so rare. Most of the business in this staffroom is conducted in an unmarked interactional context, where joking is the conventional key. There are many professional situations where this is not the case (for an example of 'no laughing matter', see Hester, 1996: 262–3), but in this staffroom there is an expectation that business and pleasure will be mixed. The teachers' views of humour indicate that this is part of the group's 'deep structure', influencing the surface structure of its interactional patterns and decision making (Frey, 1996), and it is surely not coincidental that it has developed in a back region, where the seriousness that characterises professional encounters is left behind and front region performances are knowingly contradicted.

While it is taken for granted that some situations are inherently serious, and professional encounters usually fall into this category, we perhaps also need to entertain the possibility that other professional contexts may be inherently humorous. The professional identity of these teachers is not set aside in the staffroom, nor (as many researchers have suggested) is shop talk left at the staffroom door, but humour serves to establish the necessary emotional and psychological distance between professional encounters and issues as directly experienced in the classroom, and the same matters as represented in the teachers' private space.

To understand this more fully we need to examine the ways in which humour is deployed within classroom anecdotes in order to involve colleagues in a response to a pedagogic issue. These stories focus on pedagogic incidents representing a challenge to the teacher and usually include an extended evaluation section in which audience and teller address the pedagogic issues raised. Within them, the contrast between the centripetal pull of vicarious involvement with the experience itself and the centrifugal push towards critical distance that humour establishes through its representation of the other represents an interesting professional dynamic and a potential source of creative tension in the staffroom. As we shall now see, in their representation of frontstage performance to fellow 'actors' in a backstage setting, these stories touch the very essence of life in the back region.

Stories

The power of stories and storytelling is universally recognised and, if storytelling is 'a means by which humans organize and understand the world and feel connected to each other' (Tannen, 1989: 102–3), this is nowhere more true than in the Pen staffroom. Here stories seem to hold a special place, drawing upon many of those key aspects of the collaborative culture that bind its members and represent its distinctive character. Little (1990: 515) has described stories as an 'omnipresent feature of teachers' work lives' but pointed out that 'we know little of the contribution that teachers' stories make when embedded in the wider pattern of professional interaction'. In order to address this issue and reveal how such features are woven into the texture of staffroom tales, what follows (based on Richards, 1999) works from an examination of the structure of narrative, drawing on Pen stories for illustration, to an analysis of the interactive nature of storytelling, demonstrating the special place that stories hold in the collaborative culture of the Pen school.

The analysis will draw on the well-established concept of audience as co-author, introducing the idea of *making prototypical* to represent a situation where universality can be established without sacrificing the vital element of personal experience, which is central to the success of all stories. This recognises the extent to which stories – and the realities they are designed to represent – are jointly constructed, with listeners influencing the manner of telling and even the outcome (e.g. Polanyi, 1985; Duranti, 1986; Goodwin, 1986; Mandelbaum, 1989, 1993; Ochs et al., 1992; Norrick, 1994; Mulholland, 1996) and suggests that in a collaborative professional context it is incumbent on the teller to frame his or her story in such a way as to maximise the opportunity for joint ownership.

The structure of stories

In view of the claims that will be made about Pen anecdotes (in what follows, *narrative*, *story*, *tale* and *anecdote* will be used synonymously), it is worth establishing at the outset that their structure is perfectly conventional. They certainly conform almost uncannily to Ochs et al.'s definition of a story (1992: 43) as '[n]arrative activity which articulates a central problematic event or circumstance arising in the immediate or distant past and the subsequent past, present and future actions/states relevant to interpreting and responding to the problem.' The term 'narrative activity' is important here, for, as we shall see, the act of telling may perhaps best be described a 'performance' (cf. Wolfson, 1982) with the roles of teller and audience modulated in significant ways. Interpretation is also an important aspect of Pen stories.

Structurally, they also correspond closely to Labov's widely used model for stories (Labov and Waletzky, 1967 in its first version, revised in Labov, 1972/77), with the following Pen story illustrating all aspects except the result ('what finally happened'):

Extract 5.7

```
001              (3.0)
002  Annette:  Because I just let something   Abstract
003            happen then that I shouldn't
004            have let happen. E:r (.)
005            the:: (.) they've been         Orientation
006            doing a story about-reading
007            about James Bond, and then
008            >we were-< answering some
009            questions on it, and one of
010            the questions said em::
011            (0.5)
012            er (.) >I can't remember
013            what the que- >the exactly<
014            question was but it started
015            with Bond, (.) Bond (.) did
016            such and such and they were
017            (.) to say whether it was
018            true or false.=
019    Keith:  =Right.
020  Annette:  And e:m (.) Shafie got        Complicating
021            out his dictionary and        action 1
022            was looking up (.) what I
023            thought was an important
024            word in the question that he
025            didn't understand, (.) and
026            he was looking up 'bond'.
```

```
027   Keith:      HA!┌Hahahah
028   Annette:        └**Because** it came at        *Embedded*
029                   the beginning of the            *orientation 1*
030                   s̲entence, (.) er (.)
031                  ┌he therefore          ┐=
032   Keith:         └Right right right.    ┘
033   Annette:      =didn't realise that that
034                   capital letter meant that
035                   it ┌was ┐ a n̲a̲me, (.)=
036   Keith:           └Yes.┘
037   Annette:      =**and** he's (.) he s-             *Complicating*
038                   he sho̲wed me in His              *action 2*
039                   di̲ctionary
040   Annette:      **because** I- I thought I'd       *Embedded*
041                   better go and che̲ck what         *orientation 2*
042                   he was ┌lo̲o̲king ┐ up.
043   Keith:               └Oh right.┘
044   Annette:      **And** then he said 'It's         *Complicating*
045                   this 'b̲ond', it says             *action 3*
046                   'money' and 'st̲o̲cks and
047                   sha̲res' or something.
048                   °And° l̲ots of me̲anings.'
049   Keith:        Hahaha::h=
050   Annette:      **And** I said 'No no,
051   Keith:        (xxxx)=
052   Annette:      =*it's it's*(.) Ja̲mes
053                   B̲ond,' >I mean< I po̲inted
054                   to the name on the bo̲ard
055                   and he said 'O::h ye̲s.'
056                   Heheheh
057   Keith:        Beautiful.                          *Evaluation*
058   Annette:      **But** I thought I should
059                   have p̲icked up on that
060                   earlier.
061   Keith:        It's nice though. Re̲al
062                   confusions.
063   Annette:      Yes.
064   Keith:        Yeah. Mmm. (.) Heh
065                   (3.0)
```

The story is preceded and followed by three-second pauses, which set it off clearly from the surrounding discourse, although 'because' offers a putative link with the (entirely unrelated) exchange that has just taken place. The story begins with an abstract summing up the nature of the anecdote and moves quickly to an orientation that sets the action in its lesson context. When this is acknowledged by the listener the teller introduces the complicating action which, as far as the hearer is concerned, seems to end the tale, at least judging by the outburst of laughter. The teller, however, has not completed the tale and offers further orientation to explain the significance of the dictionary search. Finally, via further orientation, the anecdote is concluded. The hearer's laughter again interrupts the telling, the conclusion of which is signalled by the teller's laughter tokens: the first, 'it's' repeated while laughing, does not generate the expected response, so another attempt is made via 'I mean', this time concluded with laughter. Although this fails to generate further laughter from the listener, it does produce a positive assessment of the story as a whole: 'Beautiful'.

Norrick has suggested that 'personal anecdotes have no punch line as such' (1994: 425), but this seems not to be the case here. The overlapping explanation, which constitutes a rejection of the laughter following 'he was looking up "bond"', is a clear indication that the climax of the story has not been reached yet, while the laughter tokens at the end of the story serve the opposite function. The punchline of the story may have misfired but it is clearly part of the design.

Other features of the telling of this tale are for the most part consistent with the surface structure identified by Labov (1972/77: 376). He suggests that conjunctions, including temporals, are to be found, and this story exhibits a remarkably consistent pattern (bold text): once the orientation is complete, the story begins with 'And', a conjunction which marks each advance in the plot, while embedded orientations are signalled by 'because' and the final evaluation, which links neatly with the opening abstract, by 'but'. The text also exhibits the simple subjects predicted by Labov, as well as preterite verbs and an underlying auxiliary, in this case a past tense marker. Labov, however, predicts a range of adverbials, of manner, time and especially location, which do not feature in this text. Neither are they a noticeable feature of other classroom stories in the data. This may be because the classroom represents a taken-for-granted setting and the focus is very much on interaction between teacher and students rather than on the specific actions of either. It does, however, point to the fact that the 'drama' in such stories lies not so much in physical as in intellectual engagement. This is something that may in part explain the extended evaluations that usually accompany them.

What also sets this apart from most other anecdotes in this Pen staffroom is the relatively passive response of the recipient. As Jefferson has shown, it is possible to respond to story telling by using 'continuers' such as 'uh huh' or 'mmm' where the 'user is proposing that his co-participant is still in the midst of some course of talk, and shall go on talking' (1983: 4). Since the 'complicating action' is so brief in this story, even these are not to be found, although the overlapping 'right right right' and 'yes' in the first embedded orientation indicate receipt, and the change of state token in the second embedded orientation acknowledges its relevance. There is in the first embedded orientation, though, an indication of audience interference: the overlapping acknowledgements are intrusive and carry at least the suggestion that this particular orientation is not necessary, that the significance of the student's looking up 'bond' has been recognised and that further explanation is redundant.

Evaluation and participation are two aspects of story telling which are never far apart. There is always a point to a tale, and the teller will seek to align the audience with his or her own evaluation of its significance. In order to do this it is necessary to interpret and respond to the audience's reception of the story, while the audience in its turn may seek to appropriate the tale to its own ends, whether these are consonant with the teller's or not. In a professional context, where so many experiences are common and where audience and teller share similar concerns, the likelihood of involvement is so much the greater. As we shall now see, it is through such involvement, dramatic, humorous and reflective, that Pen staff choose to address the challenges and condundrums which are the concrete representations of their profession's mysteries.

Openings

Perhaps because of the staffroom's spatial and temporal proximity to the classroom, Pen stories reveal an overwhelming preference for fresh anecdotes, which accounts for the fact that 80 per cent of them have a professional focus, the figure rising to almost 90 per cent if the classroom setting alone counts as professional. Brought kicking and screaming into the staffroom, these stories usually arrive still warm with the emotions of the experience they seek to represent, and in this respect they are very different from other stories discussed in the literature. In a break as short as 20 minutes stories have little time in which to go cold, but the promptness of their introduction is striking: in half of the breaks recorded, stories are introduced within 50 lines of the start, and three-quarters of the breaks see the introduction of a story within 100 lines. The pattern may not be conclusive, but it is at least indicative of a willingness to relate recent professional experiences.

In fact, as I shall now show, the professional relevance of these stories is signalled in a highly unusual way – and one which may be unique to the back region. There are two aspects that will need to be considered in order to demonstrate this: the way in which the story is introduced into ongoing conversation, i.e. its relationship to what has gone before; and the way in which the story itself is prefaced, i.e. the relationship between the announcement of the story and its subsequent development.

'Stories', says Jefferson (1978: 220), 'emerge from turn-by-turn talk, that is, are locally occasioned by it' and most of the stories in the Pen data set conform to this description, emerging as a natural part of ongoing discussion. There are various ways in which this is achieved:

1. The story is prompted by a comment made by another interactant, sometimes immediately, sometimes after a pause, as after *Lot of rain last night. I was surprised. (1.5).*
2. The storyteller responds to a direct invitation to recall individuals or incidents such as the following: *Do you remember the Thai students we had who had (.) nicknames (.) that were names of ca:rs.*
3. A story continues the theme of a preceding story: *Yes there was a boy at my primary school called Goliath.*
4. A new story keeps within a topic established either by the speaker themself or other interlocutors: *Mmm. The em:: (.) I remember a lesson something similar to that.*

What is highly unusual in this setting, however, is the presence of five openings that offer not the slightest concession to relevance, apparently violating the 'local occasioning' to which Jefferson refers:

Speaker	Previous Topic	Initiation
Annette	Serving coffee	Shadi and Chan-May have just done a brilliant em dialogue
Paul	Friends and jobs	'Mitsuko is still (1.0) a bit dreamy
Jenny	Throwing something out	(There was) an awful performance yesterday when I was showing Terri how to lock up.

| Louise | Serving coffee | I understand what e:r Paul means about the (.) getting things out of Mona because... |
| Paul | Reports and student numbers | I experienced what you told me today about Mona. |

Despite the claim that 'at times the narrative piece may be only tangentially related to the actual problem under discussion' (De Capua and Dunham, 1993: 524), I should like to argue that, in spite of appearances, these stories are in fact 'locally occasioned'. To do so it is necessary to introduce the notion of a *sanctioned topic*. In conversation, *ceteris paribus*, there are no rules relating to the significance or otherwise of any particular topic, which is why it is necessary to establish the relevance of any new topic to foregoing talk and why 'unconnected' changes of subject are often explicitly signalled as such (for a discussion of conversation as a distinct speech event and speaker rights, see Wilson, 1989). In an institutional context, there may be rules of interaction that will determine the distribution of speaker rights, the range of relevant topics etc., thus constraining the options available to participants in talk (see Drew and Heritage, 1993), but interaction during breaks is for most practical purposes indistinguishable from ordinary conversation. Although most talk is about 'shop', there are no rules concerning this, and relationships among those involved are entirely informal. Nevertheless, this is a back region in the way that, for example, the DOTS common room is not, and there seems to be at least one sense in which the professional world makes its presence felt, producing an interesting interactional hybrid revealed in the following fieldnote:

> Towards the end of this period teachers began returning to the staff room. I was already in there with Judith, both of us at desk 3, when Harry entered and began to make coffee.
>
> Paul enters: 'Well, they're beavering away, my group.' (To Harry, who has the other general group 23 hours a week.) Brief discussion about this group; Paul leaves.
>
> Jan enters: 'We didn't get into Unit 2.' (To Judith, who is sharing the 1–1 Turkish businessman with her.) Brief discussion of progress.

What's interesting about these two exchanges is the absence of preliminaries: the staff room is obviously a place where the discussion of business is legitimate and can be introduced by anyone at any stage. The above comments were not preceded by polite enquiries from the addressee or by any attempt to establish context on the part of the speaker.

It is this feature that the term *sanctioned topic* is designed to represent. Where the topic relates to shared professional concerns, such as the activity of students or the progress of classes, no interactional placement work needs to be done; the topic will be accepted immediately as of relevance to those present and appropriate responses will be forthcoming, even if these consist of no more than acknowledgement. The fact that teachers are prepared to enter the room and speak immediately upon these topics without weighing up the talk already in progress – an interactional indulgence normally extended only to young children – is an indication of the power of this particular norm. It also reflects a preference that in this data set is confined to the Pen teachers: they like their stories hot off the press.

In the light of this, the reference to named individuals in the openings of the five apparently anomalous cases is clearly significant. With only one exception they refer to students, and the exception is that of the principal, Jenny, who rarely teaches and whose sphere of concern is her colleagues and the running of the school – which makes her reference to a colleague as much a matter of relevant business as her colleagues' references to their students. It is also worth noting that the examples here cover four of the five core teachers in the Pen, which suggests that the convention is both understood and used by all. In this sense, then, these apparently uncontextualised starts are as clearly 'locally occasioned' as stories which develop from and are embedded within the surrounding talk.

Sacks has argued that stories are preceded by a story preface, which represents 'an offer to tell or a request for a chance to tell a story or joke' (1974: 340), but the idea of sanctioned topics – and the nature of recipient responses to the initiation of these – suggests that acceptance is binding. It is interesting, if mildly embarrassing, to note that the only example in the data of such a story being 'refused' occurred when I was the audience. The inappropriacy of my failure to respond supportively to one of Louise's stories and my abrupt change of subject at the first opportunity is underlined by her reintroduction of the same story later in the break, producing a follow-up story from Paul. Goodwin (1986: 298) points out that a story preface offers recipients 'key resources that they use to analyze the story as it is being told' and 'provides them with information about the type of alignment and response that is appropriate to the story.' My lack of alignment in this case arose from a failure to recognise the significance of the preface as an introduction to a 'student problem' which demands explication and analysis. As the next section will show, such alignment is routinely forthcoming among the core staff at the Pen.

Shared stories: making it real

In addition to establishing critical distance between teachers and their professional world, the potential of narrative to generate involvement also means that it can serve as a foundation for a critical evaluation of the circumstances it seeks to describe. This section will use the analysis of a typical story to explore collaborative aspects of storytelling among the Pen staff and suggest that involvement in the process of storytelling serves to reinforce personal and professional relationships. The discussion will focus on the different ways in which such involvement is established and maintained through a process of sharing generated through aspects of the storytelling process.

Involvement is to some extent a slippery concept, especially when it is identified in terms of 'involvement strategies', which may themselves be identified in terms of the 'involvement' they generate (Besnier, 1994), a situation is further complicated by the fact that there is a sense in which all conversation is based on a degree of involvement (Fais, 1994). However, as Norrick has noted, the 'the personal anecdote rates high on the scales of involvement and rapport' (1994: 428–9) and for the purposes of narrative analysis it seems legitimate to assume that where any of the audience share in the telling and/or evaluation of a story they are 'involved' in it. Whether or not laughter or 'continuers' constitute involvement is perhaps a matter for debate, but direct participation is unproblematic.

Difficulties with students represent the most common theme in Pen teachers' anecdotes and the potential for involvement begins almost as soon as Paul has finished setting up one such story by giving examples of the language point being covered:

Extract 5.8a

```
001   Paul:   Get it on the cheap. He he'd love to sort of
002           (xxxxxx) 'I'm dying to really (xxxxxx).'
003           Often to do with em (.) >things like< (.) like
004           'dy::ing' (.) to go on holiday,' or 'I'm dying to
005           go the toilet' 'I'm dying ⌐to
006   Harry:                             | Heheheh
007   Susan:                             └HehaHAH
```

```
008   Paul:   'I'm dying for a drink.' 'I'm dying (.) °for a
009           smoke.° And other stuff. And (.) for the last
010           fifteen minutes she's not with me and she's-
011           she's going
012           (1.0)
013   Paul:   and she keeps on slo:wly opening her dictionary,
014          ┌and I say┐ 'What >are you looking for?'=
015   Harry: └ hhhh hhh┘
016   Paul:   => and she says< 'Nothing!'
017           ((General laughter.))
```

In the case of the Pen, this involves dramatic representations of student talk as soon as the minimum necessary scene setting has been accomplished, made possible not only because so much shared knowledge of the classroom can be assumed but because the use of dialogue is a common form of shorthand in the staffroom (for the relevance of shared knowledge, see Ochs et al., 1992: 68). But direct speech is more than simply routine or a matter of linguistic economy; it serves as a powerful involvement mechanism, drawing the audience into the story:

> When speakers cast the words of others in dialogue, they are not reporting so much as constructing dialogue. Constructing dialogue creates involvement by both its rhythmic, sonorous effect and its internal evaluative effect. Dialogue is not a general report; it is particular, and the particular enables listeners (or readers) to create their understanding by drawing on their own history of associations. By giving voice to characters, dialogue makes story into drama and listeners into an interpreting audience to the drama. The active participation in sensemaking contributes to the creation of involvement.
>
> (Tannen, 1989: 133)

It might also be argued that the use of dialogue allows a form of direct access to the experience for the listeners. Although the words themselves may not be verbatim, they are treated as such by the listeners and provide a text for the purposes of comparison and evaluation. Given access to the live example, the audience is able to relate this to their own experience in the classroom and the professional reflections it prompts, while at the same time framing their evaluation as a response to a precise representation of an event that has been placed before them.

The dramatisation has its own dynamic and draws on a number of other performance features. Of those identified by Wolfson (1982: 24), direct speech, repetition, expressive sounds and the historic present are all to be found, as is the deliberately chosen humorous example, 'dying to go to the toilet'. The punchline comes in the form of an exchange between teacher and student and is clearly designed to produce the general laughter which follows. This formula is repeated in the next section of the story:

Extract 5.8b

```
018   Paul:   And er
019           (0.5)
020           and there's this ┌look on her face.
021   Harry:                   └You're (dying)
022           (0.5)
023   Paul:   You know, and you finish that, and I say
024           (.) 'Okay, (.) you finished a bit late so
025           (.) let's do (.) twenty minutes (longer).'
026           And then she's stood looking, really
027           confused and
028           (0.5)
029   Paul:   there's something going on in her head.
030           (0.5)
031   Paul:   And (.) I said (.) I said 'What's the
032           problem?' and she says
033   Harry:  She says she has to get out by ten
034           forty. ┌heheh hhh hh
035   Paul:          └'dy::ing to.'
036           (0.5)
037   Paul:   And I said 'Well do you understand it means
038           'want to'.' So she opens her dictionary I say
039           'What are you looking for?' >She says< 'Just
040           checking.'
041   Harry:  >Heheh< haha ┌hahahah
042   Paul:               └And I say (.) 'What don't you
043           believe me!'
044   Susan:  hhh hehhh
045   Harry:  Yeah I often say ┌that.
046   Paul:                   └'Just checking' 'Yes, am I
047           (telling youxxxx)'
048   Keith:  Yeahahah
```

This development is interesting not least because of Paul's use of the historical present in the tale, a common feature of Pen stories as of others. A spirited debate on the relationship between present and past tenses in narrative (see, for example, Wolfson, 1979, 1982; Schiffrin, 1981; Fludernik, 1991) extended interpretations beyond those focusing merely on its dramatic impact but yielded no firm conclusions. In this example it is noticeable that Paul uses the past tense for his own statements and the present for descriptions of the student's actions (e.g. ll.29–32), reserving present tense exchanges for punchlines (ll.14–16 and ll.39–43). Apart from the obvious dramatic effect of this, it also serves to create a sense of the teacher working through an orderly sequence (recalling the retrospective orderliness of past tense narratives), while the student responds unthinkingly to present stimuli, though there is no evidence to suggest deliberate design in this.

However, something even more interesting seems to be taking place here. While a number of writers have acknowledged the capacity of narrative to exploit displacement, or the ability to refer to past time (e.g. Toolan, 1989: 4) and to involve its audience, the role of the present tense in this seems to have been ignored. As long as what is being referred to occurs in a specific past time, access to it is restricted to those present at that time (usually the teller), but when it is presented in the present tense as a dramatic re-enactment, contributions from the floor do not appear as anachronistic. This is not to suggest that contributions to a tale presented in the past tense are impossible; it is rather that the *conditions for participation* are more favourable when the situation is cast in the present tense.

In the above extract, Harry's use of the present simple form of 'say' in 'I often *say that*' echoes Paul's 'And I say' (l.42), taking a particular example from the past and highlighting its typicality, so that a shift from past to present by the teller has prompted a move on the hearer's part from minimal responses to the statement of a general truth. More significantly, this sets up an observation by Harry that prompts more extended participation:

Extract 5.8c

```
049  Harry:  Could get into a real John Cleese
050          routine ┌(that way)
051  Paul:          └ 'Just checking. Yes I do
052          yes. ┌(I'm xxxxxxxxxxxxxxxxxxxxxx┐ xxx) yes=
053  Harry:       └ '°Yes° > (Is that wrong!) '<  ┘
054  Paul:  =that's right.'
055    Ed:  ┌┌Heheh
056  Harry: └└Heheh ┌HAHA:H  ┐ =
057  Susan:         └Hahahah ┘ =
```

```
058   Paul:   ='Oh I'm sorry I thought I ⌐was just'
059  Harry:                              └ 'All these
060           years and he's using it right.'
061  Susan:  Heh hahahah=
062  Harry:  ='By all means go and check it, °yeah°.'
063  Susan:  hh
064   Paul:  Em
```

Harry's opening statement here is a clear invitation to map the classroom exchanges onto those associated with the comedian John Cleese. Implicit in this is the image of suppressed rage on the part of the 'service provider' which has been generated by perceived 'customer' unreasonableness, evident here in the mode of delivery (e.g. the exclamatory 'Is that wrong!' and the stress on 'sorry' in 'Oh I'm sorry'). Ochs et al. argue that 'the shared experience of co-narration is critical to their relationships and the instantiation and ongoing reconstitution of their familiarity' (1992:68), and the contribution of these exchanges to a collaborative culture does not need to be spelt out.

The nature of the involvement here, though, serves professional as well as personal ends. The addition of amplificatory examples combines with other supportive elements, such as expressions of understanding or commonality (e.g. Harry's, 'Yeah I often say that') to produce the effect of *making prototypical* the story being presented. If members of the story's audience can add examples of their own to those being offered or can confirm features of the anecdote as in some way typical, then the story may be taken to represent a prototypical situation, so that any solutions to the problems it generates can be interpreted as general rather than individual. In other words, the generative power of the story in terms of professional relevance is thereby increased: the prototypical situation now serves as a basis for solutions that have general relevance. The importance of this becomes clear in the process of evaluation, which begins in the next stage of the story:

Extract 5.8d

```
065           (2.0)
066   Paul:  And er and (.) so I'm saying (.) 'No:!'
067  Harry:  HEHEHeheheh (.) °(xxxxxxxxxxxxx)°
068   Paul:  And er (.) I say 'Ne- next time (.) just say
069           (0.5)
070   Paul:  er (.) 'How do you use it?' But rather than
071           sa- s- she just broo:ds on these things and
```

```
072              it's so: it's awful because you can see she
073              just- you- you just lose her. For five
074              minutes.┌And ┐ we're doing fairly basic stuff=
075  Keith:          └Yeah.┘
076  Paul:   =you see- fairly ordinary ┌stuff┐ not- not at=
077  Keith:                             └Yeah.┘
078  Paul:   =all difficult. But she gets these obsessions
079              with something.
080  Harry:  Hehmm
081  Paul:   And if you point out (.) a different tense
082              to her, Her (.) >she starts
083              going< (.) hhhh ┌hhh ┐ she=
084  Harry:                     └·hhh┘
085  Paul:   =goes into (.) think mo:de.
086              ((General laughter.))
087  Paul:   °You can see it happening.°
088              (0.5)
089  Paul:   And e:r (.) you know, (.) 'I have started
090              volleyball training.'
```

A return to the main story is signalled by the reappearance of 'and' (highlighted for ease of comparison), which was missing from the 'John Cleese' side sequence, but the interesting move here is a shift in the story's focus marked by the two uses of the adversative conjunction 'but'. They are followed by more general statements which seek to represent the state of the student's mind: 'she just broods' and 'she gets these obsessions'. This feature, which seems to be a form of accounting, is something that to my knowledge has not been remarked on in the literature on storytelling but which is pervasive in those Pen anecdotes that focus on students.

This is clearly not the sort of typification identified by researchers such as Rist (1973), Hargreaves (1977) and Hammersley, (1980, 1981, 1984), where characteristics are assigned to students and allowed to represent them in the place of deeper consideration of their character and circumstances. In fact, in the Pen this accounting is often the first step in a more complex process of exploration and evaluation:

Extract 5.8e

```
091  Paul:   And you say (.) you- you- you notice that (xxxxx)
092              been starting but I've started. And she goes 'Ah
093              no.' (That's) something that just to (xxxxxx).
094              Doesn't (it xxxxxx). But it's not just with (.)
```

```
095              difficult things, it's as if she's looking for
096              things to worry about. You get quite frustrated
097              because she just (xxxxxxxx) just tell me in her
098              own way.
099  Harry:      Yeah.
100  Paul:       And then she's stuck and she's a sort of prisoner
101              (.) locked into the text.
102              (3.0)
103  Paul:       And it rea:lly does
104              (2.0)
105  Paul:       inhibit her. ⌈In ⌉ many ways, because if=
106  Harry:                   ⌊Mm ⌋
107  Paul:       =she's- (.) if she's (.) locked onto (.) 'Get
108              out of bed,' when you're basically (.) just
109              trying to say (.) 'Why does- (.) why is it
110              a problem.' And she's just so locked onto a
111              word she can't actually- (.) tell (A) what she
112              under⌈stands
113  Harry:           ⌊>It's not as if there's any difficulty
114              about-< (.) getting out⌈of  ⌈bed anyway. Getting=
115  Paul:                              ⌊No!⌋
116  Harry:      =out of bed yes that's a problem but (.) heheheh!
117  Keith:      Yeah. (.) Eh
118  Paul:       God's it's- this=
119  Harry:      =The actual phrase is⌈(xxxxxxxxxx)⌉
120  Paul:                            ⌊(xxxxxxxxxx)⌋ You just
121              think 'Oh Christ!'
122  Ed:         It's not so much that she's kind of locked
123              onto it so she's- it's as though she's kind
124              of got- she feels as like as though er-
125              em a precipice in front of her, and then
126              unless ⌈she ge⌈ts that⌉cleared.
127  Paul:              ⌊Yeah. ⌋       ⌉
128  Harry:                    ⌊ Yeah.⌋
129  Paul:       =Yeah. Yes. ⌈Yeah.  Right. ⌉Well that's it. She=
130  Ed:                     ⌊in front of her⌋
131  Ed:         =just ⌈stops.
132  Paul:             ⌊Yeah.
133  Harry:      Heheheh=
```

The shift to accounting is clear in this section, but even more noticeable is the way in which this settles on a particular image as a representation of the problem. Paul formulates the problem as one in which the student is a 'sort of prisoner (.) locked into the text' (ll.100–01), and reinforces this by the repetition of the word 'locked' (l.107). Ed, however, challenges this characterisation (l.122) and offers instead an image of the problem which presents the student as a person with 'a precipice in front of her'. This is immediately accepted by Paul and Harry, the former embracing it almost enthusiastically ('Yeah. Yes. Yeah. Right. Well that's it.').

The difference between the two images is interesting: where the student is seen as a prisoner, locked into a text, the sense of challenge generated by 'precipice' is missing and the association is with transgression rather than adventure (and associated danger). There is no interactional evidence that the teachers involved interpret the images in this way, but what does seem to be significant is the clear difference between the images themselves and, implicitly, the accounts which might be associated with them. By couching this difference in terms of an alternative image, Ed does not directly challenge the teller's interpretation; rather, his representation serves as a contribution to the story itself, part of a collaborative effort to make the situation as vividly real as possible. From this a possible solution emerges:

Extract 5.8f

```
134   Paul:  =Yeah but if you really ┌need
135    Ed:                           └And makes
136          sure (.) ┌she knows (what it is.)┐
137   Paul:           └It really is a: a       ┘terrible
138          strategy. (.) Because it's quite obvious (.)
139          >that- that-< that she can understa:nd.
140  Harry:  Yeah.=
141   Paul:  =I- if she:
142  Keith:  Yeah.
143   Paul:  If she stopped worrying about
144          It ┌she┐ could understand it.
145  Harry:     └Mm ┘
146  Harry:  Yeah.
147    Ed:  Yeah.
148   Paul:  A::nd I say you know- i- if you keep (.)
149          ┌(xxxxx eye) to a text, you're never=
150    Ed:  └(xxxxxxxxxxxx)
151   Paul:  =going to be able ┌to┐
152  Keith:                    └Ye┘ah.
153   Paul:  t- even talk about things.=
```

```
154  Harry:  =Mm=
155   Paul:  =You're never going to be able< to find a way of
156          expressing yourself.
157          (0.5)
158   Paul:  I'm always exhausted by the end of it.
159          (0.5)
160  Harry:  (xxxxxxxxxxxxx) hhh
161   Paul:  Yes. Even me.
162          (0.5)
163  Keith:  hhhheheh
```

The solution in the end seems to lie in the adoption of a different strategy on the part of the student, and the effort of trying to establish the conditions for this leave Paul 'exhausted'. His is the leading role in this final section, but the 'finale' calls for the presence of all the actors: Harry, Ed and Keith all appear on stage to offer appropriate support for the position which Paul has adopted. The play itself, though, is one of a series, and problems such as this one will be explored and analysed again and again in the Pen staffroom.

Conclusion

'A story', says Rosenholtz (1989: 125), 'is constitutive; it makes for collective identification.' Although the content and structure of Pen stories reflect this orientation, they also have an important professional contribution to make. They function, for example, as news stories, keeping colleagues up to date with developments in the progress of individual students and classes. In a small school teachers are likely to encounter at one time or another all of the students registered on courses, and exchanges of information are therefore important. Teacher stories add to shared knowledge of the student body.

The stories also offer the opportunity to explore a common interest in approaches to teaching and concerns about students who threaten the co-operative foundation upon which foreign language teaching depends. Where students are the focus of attention, such stories provide the opportunity to share the sense of frustration that arises from failure in the classroom. Ultimately, the solution to classroom problems is to be found not simply in methodological innovation but from insights into the psychological springs of individual behaviour. Stories provide the opportunity to release some of the frustration that arises from this, a chance to draw support from others who have faced similar challenges and to explore possible responses.

Finally, in a staffroom where it is important not to take oneself too seriously, teacher anecdotes are a source of amusement and diversion. Some are told simply for humorous effect, while others invoke humour as a source of comfort and solidarity. The evaluation of human conduct implicit in this humour also serves to establish important critical and psychological distance between the event and its instantiation through narrative.

Above all, though, stories serve to strengthen the interactional threads that bind the staff together. Where tellers rely on a sanctioned topic to legitimise the otherwise decontextualised introduction of their story, this reflexive act both exploits and reinforces the collective understanding of what constitutes a relevant subject. Colleagues' contributions to the dramatisation of events that follows reinforce this shared perspective, constituting not merely a form of involvement in the events themselves but a representation of these events as prototypical of the situations Pen teachers face. The jointly constructed story then becomes an affirmation of common understanding and shared professional commitment.

I have suggested in this chapter that the characteristics of the back region make possible a creative redefinition of professional experience that is achieved through humour and stories. Talking about professional encounters is the stuff of the back region, but it would be a mistake to assume that such talk is trivial simply because of the context in which it occurs: serious business does not necessarily have to be done seriously. The next chapter will expand the examination of professional talk from back region to backstage, focusing on talk about systems and procedures in order to reveal interesting similarities and remarkable differences among the three groups.

6
Seeing Things Our Way: The Establishment of Common Perspectives

Introduction

Although differences among the three groups have already emerged within the broader collaborative context so far explored, this is only to be expected: the difference between dynamic collaboration and pallid conformity depends on it. So while argument in the Pen and IEC is similarly managed, the slightly ambiguous position of the head of DOTS generates perturbations in its negotiation there, and while the academics and research scientists are happy to engage in personal banter, the teachers tend to avoid this, engaging in gentler forms of humour. We have also seen how the special status of the Pen staffroom as a back region explains certain interactional characteristics not typical of the other settings. Nevertheless, all three groups have developed ways of speaking that both reflect and reinforce their collaborative orientation. This chapter and the next examine what is distinct about the ways in which they do this, first in terms of their actions as insiders and then from the perspective of their treatment of outsiders.

Since it would be fruitless to attempt a specification of all the interactional characteristics defining a particular group, this chapter will necessarily be selective, but it will nevertheless reveal where distinctive differences are to be found. In order to do this attention will shift to the ways in which the different groups deal with the systems and procedures that permeate their professional lives. All professional groups live within institutions, and the umbilical connection between the two is established through systems and procedures. By examining how members of different groups orient to these systems, we might therefore have at least some hope of bridging the dualism between individual and institution that Stubbs (1992) identified as the analyst's most fundamental challenge, a relationship at the heart of all analysis of institutional interaction (Schegloff, 1987).

The similarities and differences that have emerged from analysis in previous chapters may have resulted from any number of factors, not least the interplay of individual member characteristics and interactional styles. This does not undermine their importance or their relevance to the ways in which collaborative groups go about their daily business and sustain themselves over time, but – to appropriate Goffman's dramaturgical metaphor – this perspective relegates the institution to a feature of the backdrop that derives incidental significance from the particular actions and positions of individual actors. An insight into how these groups deal with the systems on which their professional existence depends does not haul the institution onto the stage as merely another actor contributing to plot and character development, but it does recognise a subtler orientation to thematic realities that helps to account for particular ways of doing business.

In order to give due prominence to the particularities of setting, the following three sections will be assigned to individual groups, each section beginning with a brief summary of the part that systems and procedures play in their professional lives. This will not serve as an analytical resource, but as analysis develops the characteristics that emerge will be related to the broader context, a combination of these forming the basis for speculations on the nature of systems talk in the three institutions.

DOTS: binding procedures

The research scientists in DOTS live by procedures; their professional lives depend on them and they have been trained in the proper observance and recording of appropriate steps. That one of the meetings in the data set is given over entirely to a review of the various checking and recording procedures to ensure proper safety and quality control is no more than a reflection of this orientation, as is the fascination for the unfolding of the experimental process expressed by Penny in an interview. Their talk has developed within this context and their interactional style reflects the demands it makes. This is most clearly manifest in the regular workplan meetings where the two senior scientific officers sit down together with schedules to plan the forthcoming week's activities.

The following extract is typical of the exchanges that take place in such meetings:

Extract 6.1

```
001  Penny:  Okay. Must do's. Animal:s (.) first.
002    Sue:  Yes:. Se::u-
003          (0.8)
004  Penny:  These a::re
005          (1.0)
006    Sue:  on trial
007  Penny:  on trial
008    Sue:  all week
009          (2.5) ((Sound of pages turning.))
010  Penny:  (xxxxx) a::::::::::re in:: pens in the tunnel,
011    Sue:  In pens in the tunnel and ((name?)) must
012          start the the trial on Fri:day.
```

This seems to be a clear example of what Mercer calls *cumulative talk*, where 'speakers build on each other's contributions, add information of their own and in a mutually supportive, uncritical way construct shared knowledge and understanding' (2000: 31), or in Gordon's (2003) terms *supportive alignment*, where speakers create ties of cooperation and collaboration. The predominant features in this talk are repetition and collaborative completion.

Repetition

The extract begins with a frame/focus move (Sinclair and Coulthard, 1975) signalling a shift of topic ('Okay'), then nominating the topic in a way that reflects both the nature of the business in hand and the participants' orientation to it. Penny identifies the main category ('must do's') with its transparent claim to priority, then nominates the first topic within this. Sue's positive reception in the second line both confirms the proposed ordering and provides an example of the sort of positive responses Gordon identifies as characteristic of supportive alignment. Following these necessary preliminaries, talk on the topic itself does not begin until line 4 and all of the four spoken lines that follow are marked by repetition, first of 'on trial' and then 'in pens in the tunnel'.

The contribution of repetition to collaborative sentence building has attracted wide comment (e.g. Díaz et al., 1996, Kangasharju, 1996, Sacks, 1992, Tannen, 1984). It is, as Tannen notes, 'a limitless resource for individual creativity and interpersonal involvement' (1989: 97) with relational implications:

> repetition not only ties parts of discourse to other parts, but it bonds participants to the discourse and to each other, linking individual speakers in a conversation and in relationships.
>
> (Tannen, 1989: 51–2)

However, the repetitions do not seem to be examples of Tannen's *participatory listenership* and they do more than merely reflect understanding; in fact, they *are* the creation of the schedule, the very act of doing the work. A consideration of the second structural feature throws further light on how this is achieved.

Collaborative completion

Sue's contribution of 'on trial' in line 6 is a completion of Penny's 'These a::re', a direct response to the lengthened delivery of 'are' and the one-second pause. When this is confirmed by Penny's echoic response, Sue goes on to complete the statement that her partner had begun:

These are   on trial

        on trial   **all week.**

Completions by a single speaker have been termed *collaborative productions* (e.g. Sacks, 1992a: 144–9, 647–55) or *collaborative completions* (Lerner, 1991, 1996). Typical examples identified by Lerner in his seminal treatment (1991) are two-part formats such as if-then and when-then, three-part lists, contrast structures and preference structures, though the collaborative building in this extract is professionally rather than conventionally structured. The same applies to what Lerner (1996, 2002) has termed *choral co-production*, not evident in this extract but featuring elsewhere in this meeting, as the following example reveals:

Extract 6.2

```
001  Penny:  For week two:, we begin on Saturday.
002    Sue:  So you've got seventy two who're in on
003          Saturday,
004          (0.4)
```

```
005  Penny:  Yeah.
006          (0.2)
007   Sue:  Which will come out- (.) on:: (.) ⌈Tuesday,
008  Penny:                                   ⌊Tuesday,
```

The choral repetition in lines 7 and 8 is synchronised and represents shared completion of the statement begun by Sue:

Which will come out on   Tuesday

Tuesday

It would be tempting to interpret such intimately coordinated talk, woven tightly from repetition and shared completion, as evidence of shared thinking, but this would be illegitimate on at least two grounds. The first, as Edwards (1997: 132–3) emphasises, is that these features are part of the interactional architecture and ways of talking, not 'overt expressions of some kind of mental communion' or 'guarantees of joint intentionality' (for further discussion of the relationship between socially shared cognition and the structure of talk, see Schegloff 1991a). These features appear in all talk to different degrees and their concentration here may be remarkable but it is not constitutive of a shared cognition. Nevertheless, and keeping this important consideration clearly in mind, the talk here is collaboratively achieved and as Crook noted (1984: 145), '[i]f intersubjectivity does become a resource to support collaboration, it is because the conventions, rituals, institutions and goals of organised social life arrange that it should do so.' In fact, this way of talking is not merely the outcome of some concatenation of such elements, a way of doing work collaboratively, but the work itself, jointly instantiated by the participants.

Penny and Sue come together at the start of their meeting aware that when they leave they must take with them the completed workplans for the coming week. Neither of them has all the information necessary to complete the plan but each of them brings to the meeting relevant papers, experience and knowledge of their respective teams, in the expectation that when this is combined with that of the other participant the result will be a workable plan. Therefore their meetings talk the plans into being (see Heritage, 1984b: 290 for the concept of talking an institution into being), and the notes that are made on the page are merely the physical evidence of what has already been done verbally.

The achievement of this depends on the repetition and completion identified above. In Extract 6.1, when Sue and Penny state that the animals are on trial, this represents the first joint action: confirmation of the animals' status. Only when this part of the plan has been completed can they move to the next part, which involves consulting the information they have on the trials (l.9), confirming that the animals are currently in pens and that the trial must start on Friday. So far nothing has been written down but the scheduling is already under way because only once the relevant step is interactionally ratified can the process progress to the next.

The importance of this ratification is revealed in the way the talk is structured in order to allow spaces for possible completions, ratifications, objections, etc. Such silences are apparent in Extract 6.1 above, but more clearly detectable in a passage where Sue makes a request that might not be acceptable to Penny. Her request comes at the end of a long silence following an extended summary by Penny of the people she needs for the tasks ahead:

Extract 6.3

```
001    Sue:  Well I wouldn't mind Ja::ne
002          (1.0)
003    Sue:  or somebody, °But Jane knows what she's
004          doing° for day for my data.
005          (0.4)
006   Penny: Yeah. I'd like Jane as well for data.
007          (0.8)
008    Sue:  So I think we try and keep her free::, >I
009          mean< What she could do is trai:n
010          (0.8)
011    Sue:  Tim
012          (0.2)
013    Sue:  on the RD bag weighing, and just keep nipping
014          down and checking it's okay
015          (0.6)
016    Sue:  while she does data.
```

Sue's indirect initial request is met with a one-second pause, an attributable silence indicating a dispreferred response (Levinson, 1983: 332–7). She therefore downgrades her request from a specific person to 'somebody' (l.3) but follows this with an explanation of why Jane is suitable for the work she has in mind. Penny's response to this is positive and she adds her own request for Jane. After a further pause, Sue offers a solution to this in the form of an upshot, explaining how keeping Jane free will work within the plan they are developing. The explanation is developed over four stages explaining how Jane can do two jobs at once: she can train, can train Tim, can keep checking on his progress, and at the same time do data (ll.9–16). Between each element in the proposal there is a pause during which Penny is free to respond, so that by the time the proposal is complete it has already been agreed by both parties, who can then move to the next item. Collaborative completion and repetition do not feature here because this is a proposal from Sue and therefore the process of its development is not distributed. It is nevertheless as much a collaborative construction as anything else in these workplan meetings.

The nature of this interaction is perhaps best captured in Cooren's *collective minding* (2004), which builds on the work of Weick and Roberts (1993) and Hutchins (1995) on collective intelligence but proposes a more processual approach that would be compatible with the position Edwards takes up that rejects mental communion. Collective minding is possible when different participants contribute to a shared project that could not be completed by any one of them alone and which depends on their shared input. For Cooren, connecting the 'here and now' with the 'there and then', produces 'a form of organizational intelligence' (2004: 519) and it is the spirit of this which informs the workplan meeting, even though the 'then' is a projected future instantiated through the plan:

> Proposing a solution thus becomes a collective action – what Lerner (1992) calls an interactional achievement – by which different actors and situations are brought together and articulated in order to produce a solution.
>
> (Cooren, 2004: 537)

This collective minding is equally evident in chaired meetings, which also reflect the DOTS team's orientation to procedures and systems and, though the interaction format may be different, its checks, clarifications and confirmations are typical of participants working within established systems which need to be implemented according to recognised procedures. An example of this is provided for the purposes of comparison in the next section.

IEC: building procedures

It is hard to imagine a more contrasting relationship to systems and procedures than that which exists between DOTS and IEC. The systems in DOTS are external, imposed by the company, by law, by precedent and, perhaps most powerfully, by tradition and training, so that the idea of making things up as they go along would be anathema to the team. What would quickly see the end of DOTS as a professional unit, though, is the very lifeblood of IEC. There are university regulations governing matters academic and various reporting systems in place, but procedural matters are devolved to academic sections and the freedom to develop systems on the hoof is one that is familiar to most academics.

In fact, the position of IEC in the university and the nature of its courses strengthens this particular aspect of its work. Originally established as a specialist language service unit for what is essentially a technological university, the centre was always regarded as something outside the normal run of things and by the end of the 1980s its activities were routinely described by senior figures in the university as a 'black hole'. Nevertheless, it performed a useful function and was profitable so it was allowed to develop Masters and PhD programmes, the former delivered by distance learning as well as in-house by the end of the 1980s. Distance learning was at that time something new to the university and the international course developed by IEC was the first of its kind in the world. The approach, with its associated systems and procedures, was developed by the IEC team to a very tight deadline and like most things developed quickly from scratch it required a number of important changes and adjustments in its early years, until all the lessons learnt could be incorporated in a completely revised programme. The course has an excellent international reputation and recruits well, but at the time of the recordings distance learning competition has burgeoned and the team know that they must maintain their reputation for quality, especially in matters of support. There is also a growing recognition that the freedom built into the new programme is affecting progression, and hence income, at a time when the university is demanding ever greater contributions to the centre. The pressure to develop and refine systems is as strong as ever in an increasingly uncertain world.

In these circumstances the challenge facing the IEC team lies not in finding ways of working with procedures but in building procedures, developing and refining them in ways that will serve professional ends.

In order to do this the team need to ensure that everyone understands how current procedures work and how any new procedures will fit the range of scenarios colleagues are likely to encounter. The team therefore engage in what might be called *wavelength talk*, interaction designed to ensure that everyone in the team is tuned into the same issues and ways of handling them. Such talk, in which upshots and formulations play a key role, is marked by explicit agreement, checks, clarifications, etc. Extract 6.4 is a fairly typical example of how team members present scenarios that require an adjustment of agreed procedures. The extract occurs at the end of an extended session during which Paul has led the discussion. As pastoral tutor he has developed a new records system designed to ensure that regular telephone contact with new students is maintained and recorded, and the team has just spent considerable time ensuring that they all understand how it works. Now that all angles seem to have been covered, Paul attempts to bring the item to an end:

Extract 6.4a

```
001   Paul:  O ↑ kay (.) that's all I've got
002          to say ⌈about that.⌉
003  Helen:         ⌊The only-  ⌋((clears throat) the
004          only (.) time when that wouldn't (.) be:
005          happening is like what (.) I'm-
006          (0.4)
007  Helen:  I will be at that stage with a couple of new
008          people in Turkey.
009          (0.4)
010   Jack:  ⌈⌈Yeah. ⌉
011  Helen:  ⌊⌊Now.  ⌋So I'll be SEEing them face to face.
012   Paul:  Yeah sure.=
013  Helen:  =So we don't need to make a phone call.
014  James:  Yeah.=
015   Paul:  =No:.
```

Helen introduces her point by indicating that there will be a single exception to the scenario the team has envisaged, then casts this in terms of a situation she is about to encounter. Only after this personal scenario does she identify the problem in the form of a pair of linked upshots (ll.11 and 13): when tutors meet students face-to-face, the phone call on which the reporting depends will not take place. Both

Paul and James agree, albeit in different forms, and Jack goes on to suggest a response:

Extract 6.4b

```
015  Paul:   =No ┌:.
016  Jack:        └W'you=note it down in
017          whate ┌ver record you keep. ┐
018 Helen:        └But I can note it do  ┘wn you know,
019          see:ing: (.) this student
020          (0.4)
021 Helen:   So what we need to do then when we make a VIsi:t,
022          (0.8)
023 Helen:   rather┌than ┐ a visit re- rather than just do a=
024 James:         └Yeah ┘
025 Helen:   visit report is put on each person's file who
026          we've seen and what we've- s-said.
027 James:   Like I did in Kessburg I
028          think, ┌Kathy.
029  Jack:          └Yeah I think that's a good idea.
```

Helen takes up Jack's response (l.18), overlapping his talk with a repetition of his key point. She then provides a further upshot, this time in the form of a solution to the problem she initially raised (ll.21–6). Once this is complete James offers an example from his own experience confirming that this approach works. His check with the programme administrator, Kathy, serves also to invite her into the discussion and to allow her the chance to contribute to his representation of the experience. Meanwhile Jack confirms that this is a good idea. There are still matters of detail to resolve, however:

Extract 6.4c

```
029  Jack:          Yeah I think that's a go ┌od idea.
030  Helen:                                  └Do we file
031          it separately, ┌just so
032  Paul:                  └Could- could we (.) could
033          we ju ┌st use the same     ┐ form,
034  Helen:        └D'you think or what ┘
035  Kathy:   ┌┌Yeah I suppo┐ se so.=
036  Paul:    └└can we just  ┘
```

```
037   Paul:   j ust write  cou ld you write a short  report=
038   Jack:    Okay Yeah
039   Helen:                 cos(xxxxxxx xxxxx)
040   Paul:   of- of your meeting.
041   Helen:       Initial face to  face con tact.
042   Kathy:                                Make a note on
043           the top.
044   Paul:        Yes. Yes. If you- if you know that you're
```

The procedural matter consequent upon the agreement to make a new report concerns the filing of this report, and Helen represents this as a straightforward question to the team. Before she has actually completed her question in line 31, Paul introduces (1.32) an appeal to use the same report form for both telephone and face-to-face meetings. What follows involves a good deal of overlapping talk featuring a number of different contributions: Paul restarts his appeal (1.36), overlapped by Jack's agreement (1.38) and an inaudible explanation by Helen (1.39), then a shared construction of the final position that can be represented as follows:

```
Paul:     ... short report of the meeting

Helen:                    noting face-to-face contact

Kathy:                              Make a note on the top

Paul:                                              Yes
```

An examination of the contributions of Jack and Julie to their respective teams will highlight the distinction I wish to draw between confirming procedures and building procedures and the implications of this for the interaction order. Extract 6.5 shows a secretary (Fiona) trying to establish exactly what needs to be done to John's job description to reflect the fact that he has the responsibility for quality assurance. The process is complicated by Joe's contributions, which not only seek to clarify the situation for Fiona but establish his own understanding as a participant in the talk:

Extract 6.5

```
001  Fiona:   So what do I do about John then, do I just sort
002           of:
003           (2.5)
004  Fiona:   forward it on (.) or just-
005           (1.5)
```

```
006  Julie:  No::=well=
007  Fiona:  =add it to his
008          (1.0)
009  Fiona:  job description an::
010          (0.6)
011  Fiona:  get you to sign it.
012          (0.8)
013    Joe:  °Mmmit'll° have to go to: (.) his manager.
014  Fiona:  It would have to go to Norman wouldn't it.
015          (3.0)
016    Joe:  Just needs a: (xxxxxx) added to hi:s (.)
017          competencies doesn't it?
018  Julie:  No:::: it's to his job
019          des⌐cription. It⌐ just needs: (.) implementa-=
020    Joe:     ⌊Oh ri:ght. ⌋
021  Julie:  =(x)it- to say responsibility for
022          implementation of q m s at DOTS=
023    Joe:  Ri:ght so they won't need signing by anybody
024          (0.4)
025    Joe:  It's just you (ou⌐tline)
026  Fiona:              ⌊Just add it into the:
027          (0.8)
028  Julie:  °(It's ⌐a)°
029  Fiona:        ⌊Doesn't his job description need
030          SIgning.
031          (1.0)
032  Fiona:  If you see what I ⌐mea:n.
033    Joe:                  ⌊°I don't know°
034          (0.6)
035  Julie:  Yeah when they're updated they have to °be
036          signed°.
037  Fiona:  So it would have to be si:gned.
038    Joe:  °e: ⌐:m°
039  Fiona:     ⌊I'd have to-
040          (0.8)
041  Fiona:  get him to sign it an:d (.) and Norman as well.
```

While there are superficial similarities between the contributions of Julie and Jack, the ways in which they move the talk forward is markedly different. When Jack (ll.16–17) recommends that the tutor 'note it down in whatever record you keep', Helen picks up on this and presents an upshot (l.21): 'so what we need to do then ...'. Jack says he thinks this is a good idea (l.29), but when Paul challenges this and proposes that they use the same form (i.e. the telephone form, not 'whatever record'), Jack is equally happy to reverse his position to provide a basis for the final jointly constructed position summarised above. Jack, then, is orienting to the suggestions of colleagues, not taking up a fixed position with regard to what is possible but ready to entertain any reasonable suggestion, even where this is different from his own. The shared position that emerges from this is the product of the team's exploration of what might be possible.

Julie, however, orients to the procedures laid down by the company and acts as interpreter of these for her colleagues. She has only three turns. The first (l.6), a rejection of Fiona's suggestion that it be 'forwarded on', prompts Fiona to propose an alternative course of action. She then rejects Joe's suggestion that all that is required is an addition to the list of competencies: 'No it's to his job description. It just needs implementa-(x)it- to say responsibility for implementation of qms at DOTS' (ll.18–22, simplified). Joe's overlapping 'Oh right' with its change-of-state token indicates receipt of the new information and on completion of Julie's point he produces a confirmation and an upshot: 'Right so they won't need signing by anybody' (l.23). Julie's final contribution, 'Yeah when they're updated they have to be signed' (ll.35–6) is repeated by Helen, who then checks on who signs it. Each of Julie's turns is prefaced with a 'yes' (once) or 'no' (twice) and there is no indication in the talk that these are anything other than categorical. In fact, the dispreferred 'no', presented baldly on-record with no mitigation, is a potentially serious face-threatening act (for a brief discussion of face see Chapter 7), but since Julie here is the mouthpiece for procedures that are not open to debate, its force is diminished because there is at least a sense in which the responsibility for it is not hers.

The paradox of externally imposed procedures is that while they constrain those who must operate within them, they also free such individuals from the pressures of time and uncertainty generated by a more opaque and less secure system. The issue here is not one of pros and cons but of how participants go about their professional business in the environments that these conditions generate. We have seen something of the way collective minding works in DOTS, but in the less ordered context of IEC the same options are not available. This is also an uncertain world but for other reasons. The team's position in the university, for example, is anomalous and members regard themselves as outsiders, partly because what they do is not understood and partly because their career profiles are very different from those of other academics. In addition, as their longest serving team member once observed, there has probably never been a time when the possibility of closure or dissolution has not been somewhere in the background. The team and its achievements are therefore a relevant matter in all their talk and, apart from a Christmas lunch and the occasional celebratory get-together, their weekly meetings represent the only realistic opportunity for team building. Unlike the other two groups, they have no common room which might offer opportunities for social or professional bonding.

These reasons may account at least in part for some interesting aspects of the team's interaction that are not to be found in the other two groups. For example, one of the features of their talk that stands out is the amount and extent of explicit agreement that is to be found. Agreement is, of course, a feature of all talk, but here it seems to be patterned into the talk so that particular clusters emerge prominently. In Extract 6.6a, for example, Pip has reported on proposed changes to teacher training courses, the relationship between these and the Masters being particularly important given developments in the latter. The talk is essentially an exchange between Jack and Pip, with the others, who do not teach on these courses and are not affected by the changes, listening for information. There is agreement throughout the talk but it emerges in chorus at three points over just 62 lines, each instance drawing in others from the team. In the first example, Jack explicitly asks for agreement and gets it:

Extract 6.6a

```
001    Jack:  Well could we agree that principles are going
002           to be dropped
003           (0.6)
004    Jack:  >because I think ┌they< should.
005    Paul:                   └Mmm ┌m
006     Pip:                        └Yeah=
007   James:  =Yeah.
008     Pip:  Yea ┌h.
009    Paul:      └°mmm°
010    Jack:  Ok ↓ a:y.=
```

Paul, Pip and James all agree to the change, both Pip and Paul repeating their assent. Since agreement has been requested, the contribution from listeners is understandable, but it occurs again 25 lines later when Jack sums up an option:

Extract 6.6b

```
035   Jack:   And it'll either be tefl
036           (0.4)
037   Jack:   [[or t] es- teesol=
038    Pip:   [[mmm]
039    Pip:   =°mmm°=
040   Jack:   =and that will just be (.) decided.
041    Pip:   Uhuh
042   Paul:   °mm°
043   Jack:   Perhaps test the waters with-
044   Paul:   Mm [mm
045   Jack:      [the:
046           (0.4)
047   Jack:   (new one)
048    Pip:   Yeah.
049           (0.4)
050    Pip:   Oka:y
051   Jack:   Good.
052  Helen:   Right.=
```

This time Paul and Pip signal assent as Jack articulates his position and when it concludes Pip upgrades her agreement by lexical-
ising it (l.48). At this point the item, and indeed the topic, is effectively closed, so Pip's reformulation of her agreement (l.50) following
almost a half-second pause is interesting. Positioned after the pause,
it draws attention to the silence that precedes it and attracts an
immediate positive evaluation from Jack, supported by Helen. In
drawing Jack back into the talk, it produces explicit confirmation
of the position from him and at the same time initiates a three-
part sequence. The significance of such sequences has already been
mentioned and it is interesting to note the frequency of its occurrence
in these environments. The first example involved a double three-part
sequence (Paul–Pip–James, Pip–Paul–Jack) and the final one illustrates a
variation:

Extract 6.6c

```
053  Tony:   =So we then c- (.) >I'm sorry<
054          will [we th-] >will we< THen have an introductory
055   Pip:        [Yes a-]
056          (1.0)
```

```
057 Tony:  certificate, an advanced certificate a
058        diploma and a masters.
059 Jack:  Yeah.
060  Pip:  Exactly.
061 Tony:  Yeah?
062  Pip:  Yes.
```

Tony's formulation of the situation summarises the courses that will be available in the new suite of programmes and receives immediate confirmation from both of relevant parties, Jack and Pip, but he emphatically invites a third confirmation, provided by Pip. The importance of formulations for shared accountability is the main focus of Heritage and Watson's seminal paper on the subject, in which they identify the following characteristic:

> Insofar as a routinely produced formulations (*sic*) may be judged by a recipient to be defective or nonpreferred, it may occasion a retrospective inspection of the conversational materials which is addressed to the multiple accountability of such materials.
>
> (1979: 138)

Tony's invitation of the third confirmation therefore counts as a final ratification – effectively the third part of an incomplete sequence – of the soundness of his formulation and hence of the preceding talk.

It is important to emphasise that these three-part sequences do not occur in the same way in the other settings and they should be seen as part of a pattern of explicit shared ratifications of the team's activities. Holmes and Marra (2004), basing their work on Fletcher (1999), draw on the concept of *creating team*, which involves social talk, humour and anecdotes, all of which are to be found in IEC talk. However, in their data explicit approval is 'off-record' and praise is often regarded with suspicion. They explain this in terms of a culture in which 'tall poppies' are not tolerated, but the same might be said of English culture. In fact, what happens in IEC is perfectly consistent with their findings because in Holmes and Marra's data approval and praise are directed at *individuals*, so 'off-record' talk has to be delivered away from the group of which these are members. In IEC, however, the praise is directed at the activities and achievements of *the group as a whole*, even where this is represented by individual members or even students. 'Off-record' here means, in effect, 'not in front of other groups'. Hence we find behaviours such as explicit praise and thanks directed at a member of the group, as in the case where Mark has just presented the proofs of a new course brochure on which he has been working and Jack draws this part of the meeting to an end:

Extract 6.7

```
001   Jack:   OKA:y yeah >all good< stuff, thanks Mark.
002           Newsletter,=
003   Helen:  =And thank you very much for
004           ge┌tting it all (xxxxx) it┐ was really┐ REally=
005   Jack:     └YEa:h ┌cracking stuff. ┘          │
006   Tony:          └Yeah tremendous thank you. ┘
007   Helen:  =go┌od I'm SO pleased.
008   Jack:      └°really good.°
```

There is even an example of noisy celebration:

Extract 6.8

```
001   Jack:   An:d (.) Tony can announce (.) our first two
002           year completed
003           (0.4)
004   Tony:   Ye ┌s.
005   Kathy:     └Ye::┌::s.
006   Jack:           └Sam ↑ Taylor.
007   Tony:   That's right.
008           (0.6)
009   Jack:   With
010           (0.4)
011   Tony:   He's done it with an A:.
012           (0.4)
013   Jack:   With an ┌A.
014   Lisa:           └'ra:::┌y
015   Helen:                 └RA┌a::::::::::y
016   Kathy:                    └ra:::y=he's┌(xxxx) an (xxxx)
017                                         └((Sound of
018           Jack ┌banging desk.))┐
019   Helen:       └Em Tina Hopper,┘ registe:red (.) in April,
020           (1.0)
```

```
021  Julie:  So she's got to
022           complete ┌by
023  Jack:           └Oh she can be the ┌second.
024  Lisa:                              └SHE WILL.
025          (0.2)
026  Lisa:  She will.
```

Notice here the almost fairground presentation of the feat. Jack begins with a pre-announcement calling attention to the item and, by framing it as an announcement, its importance. His announcement of the name emerges from Kathy's enthusiastic 'Ye::::s', and by pausing to allow Tony to present the highlight (1.10), he sets up the basis for his own a repetition of this (1.13), which increases its force. As with a successful party conference announcement, the celebrations overlap with the end of this, but there is more to come. Helen points out that Tina Hopper may also complete and Lisa, who led the initial cheering, responds with a loud, emphatic and repeated assertion (ll.24 and 26), almost in the vein of ecstatic responses in revivalist meetings. In the context of a business meeting, these responses may seem rather febrile, but they take place in the privacy of the group and represent shared confirmation and celebration of achievements gained through materials and procedures the team have developed together.

Pen: systems in place

The freedom to create systems enjoyed – or endured – by IEC colleagues is also available to teachers in the Pen, but they have less need of it. Theirs is a stable system, developed in the light of a shared educational philosophy and with no outside interference. Were it not for the demands of external inspection from two different bodies, there would be no need to take account of any procedures other than those they have developed themselves, and it is in summarising the views of an external inspector that Kate – as owner of the school, an outsider to the main group – sums up the position of the group as a whole (text simplified): 'She's very keen on staff meetings and interaction, (communication) between teachers, that sort of thing, not so interested in formal systems like Languacentre had. So I think we should suit her very well indeed.'

The Pen team can afford to take a fairly laid back view of systems and deal with them either in passing or from a fairly reflective distance. When Louise is talking about a tutorial with a student, for example, and explains that she 'threw the cat among the pigeons' by informing the student there would be no tutorial, Paul notes the reason for this (simplified): 'We never really established how we should do this and maybe we should formalise it.' However, no similar incidents occur that term and the issue of formal arrangements is left undisturbed.

In the 15 months I spent at the Pen I came across only one extended discussion that centred on systems and this has already been examined briefly in Chapter 3. However, in that chapter relatively little was made of its most interesting feature: the way in which metaphor is deployed in the discussion and the extent to which it contributes to the representation and reconciliation of the different positions taken up.

In a weekly meeting, when Harry introduces the topic of pooling teaching materials, Annette proposes something Ed had suggested the previous day: 'a system of e::r (0.5) putting photocopie:s (.) in a: big box or something.' In an extended turn, Ed points to what he takes to be the main drawback of his proposal: that they work mainly from textbooks and there are plenty of copies to go round. This is immediately followed by a representation of the situation around which the argument will eventually crystallise:

Extract 6.9a

```
001   Paul:   I- I think what we- a- a good way of looking at
002           it would be: (.) >would be< maybe different
003           strategies with materials. 'What I do with
004           this piece of material.'
005           (0.5)
006   Paul:   E:m
007           (0.5)
008   Paul:   or 'What I do with this kind of idea, I think,
009           you can imagine what that box is going to
010           look like after (.) two months if we actually
011           start putting in it. You know
012   Keith:  HEHEha!
013   Paul:   It's just going to be a heap of things. SO I- I
014           thought it- it might be an idea just to-
015           (0.5)
016   Paul:   just to look at different ways of exploiting
017           a piece of material which
018           (1.0)
019   Paul:   which we may know about, and sort of pooling
020           ideas rather than pooling pieces of paper.
```

When Paul takes the floor it is in order to disagree with the box idea, although his disagreement emerges clearly only at the end of his extended turn and is prefaced with hesitation markers, hedging and evidence in support of his eventual position. On one level the assertion about the state of the box, which is presented in the form of an appeal for his colleagues to visualise the results themselves (ll.9–13), is the justification for his claim that it is better to pool ideas than materials, but a more subtle positioning emerges from his formulation, one which appeals to the shared commitment to an active rather than passive orientation in their work.

This active/passive contrast is at the heart of his formulation, which begins emphatically and ends with a subtle transformation. The emphasis on the active 'do' in 'What I do with this piece of material' is underlined by the immediate repetition of the phrase 'what I do' and contrasted with the passive 'heap' which will result from putting materials into a box (l.13). At the end of his turn he takes Harry's opening description and divides it into two parts which reflect the division he has now established: the alternative to his suggested active and purposeful exploitation of a 'piece of material' is the relatively passive pooling of 'pieces of paper'. In teaching terms, a piece of material is something which is chosen and used, while a piece of paper is simply an object.

The contrast is not immediately taken up as Annette and Harry develop a compromise position which recognises the worth of some pieces of paper. This in turn prompts discussion of a similar system and the importance of someone keeping it up to date, something that may be problematic in the Pen, where the 'post' of academic manager (in reality a rotating responsibility with an attendant timetable allowance) has recently been abolished. At the end of this Paul reformulates his position and returns to his original metaphor:

Extract 6.9b

```
001      Paul:  Yeah. And I think that is a sort of a danger
002             >you just end up with a sump=
003     Harry:  =Yeah=
004      Paul:  =of material,
005   Annette:  Yeah=
006      Paul:  =which people say 'Oh this looks alright,
007             I'll try this.'
```

Here the 'heap' has transmuted into a 'sump', suggesting waste products and therefore a dirty pile of useless material, an image that both Harry and Annette seem to accept. However, they are quick to develop a response to it showing how an active approach will counter the problem: 'to do it properly you need to provide some sort of teaching notes'. Paul recognises the force of this and Annette's suggestion that they also need 'regular sharing ideas sessions' that will complement the teacher's notes, and he goes on to conceptualise this explicitly as dynamic. He contrasts this with a 'boxful of ideas' and it is left to Harry to represent the alternative in passive terms: as 'a load of stuff that nobody actually ever uses':

Extract 6.9c

```
001       Paul:  But (.) I mean (.) if- if we say to somebody:,
002              (.) to the British Council, we have regular
003              >let's say< twi- two every_fortnight, we have
004              an (xxxxxxxx) sharing materials,=
005   Annette:  =Mmm
006       Paul:  if it's minuted al:so,=
007   Annette:  =Mmm
008       Paul:  then that I think is (.) it shows a li- a
009              dynamic thing rather than=
010   Annette:  =Yes=
011       Paul:  ='Oh this is our boxful of ideas.'=
012   Annette:  =Yes and an on:going=
013       Paul:  =Yeah=
014   Annette:  =process.
015       Paul:  And I think ⌈they ⌉ should be satisfied.
016   Annette:              ⌊Yeah.⌋
017      Harry  I think you're right.⌈You cou⌉ld easily end up=
018   Annette:                        ⌊Yeah  ⌋
019      Harry:  =with a load of stuff that nobody actually
020              ever uses.
```

Now that a compromise position has been reached, the team reprise the issues raised and discuss possible refinements to the system. Particularly interesting is the way that throughout the discussion team members orient to the basic distinction Paul has just articulated. The stimulus for this is usually the mention of a box and the image of a heap or pile figures prominently:

Extract 6.9d

```
001       Paul:  there's something about< boxes (.) that
002              you put things ⌈in
003   Annette:                  ⌊>Yes but< if the academic
```

```
004                 manager ⌈knew that this was part of ⌉ his=
005     Harry:              ⌊(If we're going to do that)  ⌋
006     Annette:    =⌈or her job              ⌉
007     Harry:       ⌊we'd have sev⌋eral boxes down there.
008                 (0.5)
009     Annette:    Yes. Just get things piled in
```

At other times the image is closer to that of a sump:

Extract 6.9e

```
001     Annette:    But if- if ⌈we did use  ⌉ this box this would be:
002     Paul:                  ⌊When) I had- ⌋
003                 (0.5)
004     Annette:    It should be emptied fortnightly.
005                 (1.0)
006     Annette:    Because those things should be: (.) fi:led or
007                 discussed (.) or both. A- at the meeting.
008      Harry:     (xxxxxxxxxxxxxxx) discover all sorts of
009                 things in the bottom of it.
```

Eventually, Harry and Paul light on an image of materials that carries with it associations of creativity and of active processes:

Extract 6.9f

```
001      Paul:    =that's just materials though, isn't it?
002               (1.5)
003     Harry:    Like raw material.
004      Paul:    Raw material.
005       Ed:     Mmm. ⌈>You know you could just-<
006     Annette:       ⌊Yes
```

Finally, the creative process is established as the primary element in the system they have been discussing. 'Ideas' says Paul 'is better than (xxxxx) things', capturing Harry's essential point. 'Yeah', responds the latter, 'That's what I meant.' By the end of this long discussion (well over a thousand lines of transcript) Paul's initial characterisation of the issue as one between passive heaps and active engagement has found its way into the talk of the other participants and provided the basis for a shared position. Metaphor in Pen talk seems to cut across team alignments that may develop during the course of an argument and undermines the *building alliances* (Kangasharju, 2002) that would generate a more oppositional format. But it is even more pervasive than this.

In his discussion of the notion of *social representations* Moscovici (1984) argues that groups rely on shared images and ideas and that these become the cognitive context within which members communicate. There is strong evidence for this in the Pen data and the extent to which metaphor serves these teachers' professional ends is what distinguishes their talk from that of the other two groups. Metaphor features prominently in certain forms of IEC talk (for a discussion of this, see Mann, 2002, Chapter 7) but is not deployed strategically and does not cluster around two oppositional concepts, while in the procedural orientation of DOTS, where clarity and precision are important, it is hardly to be found (though, as Keller, 1995 has shown, this should not be taken to suggest that biology as a professional field is somehow resistant to it).

An illustration of the extent to which the active/passive orientation informs professional discussion in the Pen can be gleaned from a consideration of the contexts in which it appears, all of the following examples occurring within a few months of the example above:

Staff meetings

The discussion of systems above takes place in a staff meeting, but it is not the only case where the active/passive contrasts serves as the basis for developing a position. At a later meeting Paul, Annette and Louise become involved in a discussion of teaching methodology in which 'learning a list' is opposed to 'doing an activity'. Considerable emphasis is placed on the need to 'confront meaning', 'do the task', 'use' and 'search' in the important process of 'developing a conversation'.

Breaktime talk

What begins as an essentially linguistic discussion in the break soon develops into a shared attack by Harry and Paul on 'just learning stock phrases' as a substitute for a more considered approach to letter writing, and Harry's reference to 'boxes of useful expressions' echoes the 'boxful' of materials that featured so prominently in the discussion analysed above. Boxes, like lists, heaps and piles, are characteristic of a passive approach to language learning that depends on the unthinking absorption of information at the expense of 'confronting meaning' and engaging in interaction.

Preparation days

Discussions of materials are quite common in staffroom talk and what emerges from them is a belief in the power of creativity and an associated preference for authenticity and communicative potential. In an extended discussion during a course preparation day, the importance of choosing stimulating texts that can be exploited for their 'intrinsic worth' is repeatedly stressed.

Stories

The power of imagery in characterisation emerges clearly from an examination of the images occurring in the three long tales in the data set, which can be summarised as follows (arrows stand for 'refers to'):

Paul	a sort of prisoner, locked onto a word	→ Student
Ed	facing a precipice	→ Student
Paul	wading through mud	→ Teacher
Paul	going limp	→ Student
Louise	generating a bit of a spark	→ Teacher
Paul	lets it dissolve around the teacher	→ Student
Paul	like putting plastic next to a hot fire	→ Student
Paul	like jelly	→ Student

The clear division between active and passive images is apparent here. In the first story the student is represented as someone trapped, unable to make progress, and the teachers discuss possible reasons for this. The image of the teacher in the second story is that of someone attempting to move forward ('wading') but held back by the nature of the class ('mud'), or students who 'go limp'. Finally, in the third story the contrast between teacher and student is brought to the surface in terms of the effects of the active element on the passive: the teacher generates a 'spark' which produces fire, in the presence of which the jelly-like or 'plastic' student melts, causing the lesson to 'dissolve'.

The importance of image and metaphor has been recognised both generally (e.g. Lakoff and Johnson, 1980) and in the context of education and language teaching (e.g. Cameron, 2003; Cameron and Low, 1999; Hannay, 1996; Thornbury, 1991), and its presence here is part of a bigger picture. It is impossible to say whether or not this particular feature would have emerged had these teachers not worked together for so long and come to share a common vision of their professional responsibilities, but this is of only passing interest. What emerges from their daily talk is something that both reflects and reinforces a shared perspective on what is positive and what is negative in pedagogic experience. The imagery that dominates staffroom talk is simple, powerful, and pervasive, even crude, but what it lacks in subtlety it makes up for in significance. Time and again the polar images of activity and passivity emerge as points of orientation, intimately connected with the staff's emphasis on engagement and communication, and their commitment to the collaborative enterprise which is the Pen School.

Conclusion

This chapter has focused on talk related to systems in the three research sites, and the result is a surprisingly diverse collection of characteristics, from collective minding realised through collaboratively constructed representations, through explicit expressions of group solidarity, to the strategic deployment of metaphors representing core professional beliefs. In view of 'the elliptical and impenetrable nature of the language of individuals who know each other well and who form a social group' (Cutting, 1999: 179), this fairly crude sketch of obvious differences is perhaps as much as can be achieved, but it does at least draw attention to the different ways in which collaboration is established.

Establishing relationships through work talk is not, of course, the same thing as explicitly relational talk, though the latter, as Koester (2004) has shown, is pervasive and its placement extremely flexible. It has not featured in this chapter because I have chosen instead to highlight the more general characteristics of these groups. Neither have I focused on particular patterns across the groups as in earlier chapters though, as Drew observes, such comparative analysis may be necessary in order to establish which patterns are in fact generic:

> ...*comparative analysis* may be required in order to assess how far a certain pattern, device or practice is generic to talk-in-interaction, and therefore not restricted to any one type or setting; or whether, perhaps, there are systematic variations in the occurrence, scope, properties and form of certain practices – variations associated with the specific settings in which they occur and the activities in which participants are engaged in those settings.
>
> (Drew, 2003: 294)

The interactional characteristics discussed here have developed over time in response to particular histories, professional demands, local contingencies and a host of other factors, not least the strange chemistry of interpersonal engagement, and it would be interesting to trace their development. Cutting (1999, 2000) has attempted this but her group of Masters students who have come together for just one year is very different and the features she concentrates on are those that can meaningfully be measured over time. Nevertheless, her work demonstrates that longitudinal examinations of the development of talk are possible and there is every reason to believe that they would be fruitful. In the meantime we are left with interesting cross-sections providing strong support for Cooren's (2004: 543) call for a shift of focus from decision-making to sense-making. In the next chapter, however, the process of sense-making in these three settings will be exposed in a less favourable light.

7
Us and Them: Constructing the Other

Introduction

People sometimes choose to define themselves in terms of who they are not, along the lines of Samuel Beckett's assertion, 'I am not British. On the contrary.' This is an extreme reflection of the necessary truth that the existence of groups is dependent at least to some degree on the extent to which they can be distinguished from other groups and collections of individuals. So far I have concentrated on interactional features that characterise my chosen professional groups and dealt only incidentally with their relationship with outsiders. However, this relationship has been much studied from a psychological perspective because of the extent to which psychologically salient groups form a shared representation of their own identity in terms of how they differ from outsiders or other groups (e.g. Tindale et al., 2001).

In this chapter my attention will be directed towards the ways in which this shared representation is interactionally constructed, with the aim of providing a flavour of the many ways this is achieved. Unless indicated, the examples here are not the only ones of their kind, and the list is far from exhaustive. There are, for example, many ways in which the term 'we' is deployed, as the analysis of argument in Chapter 3 shows. In addition, I have excluded examples where the wider organisation is simply 'slagged off', a common feature of work talk and one that certainly generates involvement but where criticism is made explicit rather than embedded within other discussion. The aim of the chapter is not to develop a typology or even a description of how relations with outsiders function but to provide an illustration of how members of the group construct their interaction so as to invoke external agencies (individuals, organisations, etc.) as accountable participants within their discourse. In order to do this it will look at how outsiders are treated when interacting directly with the group and how they are represented when absent. First, though, I offer two brief examples where reference to outside agencies allows the group explicitly to define itself and its activities.

Who we are

The only example in the data of a group explicitly defining itself is an unusual one which recalls the ambiguous position held by John, team leader at DOTS. In this short exchange the team are yet again discussing the new personal appraisal system and John is explaining how this will involve consultation outside the group:

Extract 7.1

```
001   John:   One of the- one of the ma:jor changes is wha-
002           is really what they're calling enhanced
003           standards assurance.
004           (2.0)
005   John:   An' this means
006           (0.2)
007   John:   e::m (.) that there will now be consulta:tion
008           outSIde of the group.
009           (1.0)
010   John    [[(Both) ] Sorry
011  Maria:   [[°(xxx)°]
012  Maria:   Which group
013           (0.8)
014  Maria:   °is the group.°
015   John:   Well
016           (0.2)
017   John:   for example here.
018           (0.6)
019  Penny:   Right.
020   John:   We w-=
021  Julie:   =Outside the tea:m.=
022   John:   Outside the team if you want
```

This is a repair sequence whose trajectory begins with a trouble source in line 8, where John's use of 'group' is not clear, at least to Maria, who initiates the repair in line 11, a turn that itself has to be repaired in an insertion sequence starting with John's 'sorry' (ll.10). Judging by John's response of 'here' (l.17), he interprets the ambiguity as deriving from the fact that 'group' may refer to the organisation as a whole (as in a group of companies) or the specific group in DOTS. The use of 'for example' makes it clear that 'here' refers to this particular group (DOTS) as opposed to any other group in the organisation, hence addressing the distinction that is at the heart of the problem. Although there is no response from Maria, Penny accepts the clarification. However, when John attempts to continue, Julie provides a second repair of the original trouble, which involves a reformulation of John's 'outside the group' (l.21) by changing 'group' to 'team'. The distinction here seems to be between the organisation as a group and DOTS as a team, i.e. an identifiable group of individuals working together with common purpose, though since there are at least two distinctive teams within DOTS itself it could be argued that this shifts the ambiguity from that between the organisation and its parts to that between DOTS and its parts. However, in the context of exchange, the distinction is clear and it serves the additional end of reinforcing the group's identity in terms of common purpose.

The distinction here does not seem to arise from any attachment on Julie's part to this particular term; in fact, just a couple of minutes earlier she had raised no objection to Sue's use of the term 'staff': 'Just when you think they're getting somewhe:re by- (0.6) starting to pu̲t in pa̲:y progre̲ssio:::n, (0.6) em (.) you think (.) well maybe they're li̲stening to the sta̲:ff you kno:w (0.6) and then >they go and do something< like tha̲:t.' The term 'staff', like 'group' and 'team' are terms that are available for strategic deployment according to whichever alignment best serves the interlocutor's purpose. Hence, when Sue absorbs DOTS into 'the staff', she is identifying with a much larger constituency which is being ignored by the top management. Her objection is therefore delivered not from the perspective of a small group whose views the management has ignored (perhaps for legitimate reasons), but on behalf of all staff in the organisation; and from this perspective a refusal to listen is almost by definition a failure of management.

John's response (l.22) to Julie's alternative formulation in Extract 7.1 grants her repair but at the expense of downgrading its force: his concessive 'if you want' treats 'group' and 'team' as co-terminous. After a pause, he resumes his presentation of the situation and the issue is sidelined, but it does serve as an interesting glimpse of the ways in which a group may be sensitive to terms used to describe itself.

The second example, from IEC, provides an interesting illustration of how the group is defined in terms of one aspect of its work, and how this is used by the speaker to underline the value of that work at the expense of different approaches:

Extract 7.2

```
001   Tony:   There is QUAntitative data here,
002           (1.5)
003   Tony:   that the- >I mean< the manipulation of which
004           some people actually refer to as re ↓SEARCH.
005           (0.8)
006   Helen:  Mmm=
007   Julie:  =Yeah
008   Jack:   Yeah.
009   Tony:   Yeah for most people research is NOt-
010           (0.6)
011   Tony:   striving to understAnd in the way that we
012           unde- (.) DO:, it's a question of >you know<
013           well (.) this happened, >this happened< 'n
014           that happened.
015   Paul:   °Mm!°
```

The group has been discussing the results of a student survey and the implications of this for planning, but in representing this as quantitative data, Tony recategorises it so that it now belongs to the research rather than the planning process. In the absence of either rejection of or support for this recategorisation, he then represents an approach to research that would use such data. His represent-ation of the process (ll.3–4), which involves the 'manipulation' of data, sets up his evaluation of this as not genuine research but 'what some people actually refer to as research'. That they have no right to do so is implicit in the use of the term 'actually'. His representa-tion receives a three-part ratification (ll.6–8) and he goes on to make explicit the difference between the group's approach at that of 'most' people. The group's 'striving to understand' is set against a mechanical approach that is reducible to 'this happened, this happened and that happened'. By downgrading other research to unthinking manipulation and crude causality, Tony is able to bolster the value of the group's own research.

This reinforcement of the validity of the group's actions depends on an objectively unsustainable representation of a proposed alternative built on an interesting shift from 'some people' who falsely describe their work as research to 'most people' who adopt a simplistic approach. One of Tony's common complaints is that his own research is seen as 'soft' and is not appreciated either by the university or the research community as a whole, and colleagues are sympathetic to his position. The division he refers to is commonly summed up as that between qualitative and quantitative approaches to research or differences between neo-positivist and constructivist or critical paradigms, and the debate has at times been almost poisonous. What is interesting here is the way in which the group is prepared to go along with Tony's description of the alternatives, even though they know that his position rests on a misleading representation. In fact, although most people probably do hold this naïve view of research, this does not apply to researchers in the tradition he is attacking, as both he and his colleagues know. Nevertheless, as Holmes notes (2000b: 168), shared criticism of outsiders can 'serve to cement solidarity between work colleagues. A criticism endorsed by others reflects common values and attitudes.'

Responding to the outsider

Misleading representations of general positions are one thing, but challenging an outsider face to face is quite another. Where this does occur the challenge is either humorously framed or embedded in argument, as illustrated in the following examples.

The Pen school itself is non-hierarchical and effectively independent as far as the day-to-day running is concerned, but it is part of a larger organisation with a defined hierarchy. Kate is sympathetic to the ethos of the school but she is also, by virtue of being head of the organisation, in a position of authority. She has extensive experience in TESOL and an energetic and forthright personality, characteristics that form the basis of her engagement with the staff of the Pen, which is generally symmetrical and unproblematic. In the meeting of which the following extract forms a part she is clearly a dominant figure, but this needs to be interpreted in the light of the fact that she has been specifically invited as someone with specialist knowledge of how to prepare for a British Council inspection of the sort the Pen now faces. In this role her contribution is accepted and valued, but when she attempts to pull rank humour is immediately deployed in order to re-establish the collaborative norms that underpin interaction within the group:

Extract 7.3

```
001   Kate:  You put bits about your liaising with
002          Rockingwell [a sister school] and things.=
003  Jenny:  =Mmm
004   Kate:  Well just-=
005  Jenny:  =Mmm=
006   Kate:  =think about it will you.
007   Paul:  hh hh ┌hh hh heh
008  Jenny:        └heh heh heh ((Suppressed laughter.))
009  Jenny:  Yes miss.=
010   Paul:  =Come on. >Which ┌one of you< forgot that.=
011   Kate:                   └Mmm
012   Paul:  Hahh ┌heheheh
013   Kate:       └>What's that-<
014  Jenny:  Have you forgotten that you (had a xxxxxxxx).
015          You've forgotten to send us any students,
016          °haven't you.°=
017   Kate:  =Mmm
```

Kate's 'Well just think about it will you' is peremptorily delivered, emerging as an injunction rather than a suggestion, and it receives a response from Paul that, given the absence of any indication in Kate's delivery that it is designed to be interpreted as a candidate laughable, seems to be a clear case of laughing at rather than laughing with (Glenn, 2003: 112–13). The incongruity of the response is underlined when Jenny picks it up and attempts to suppress her laughter. Jenny's exaggeratedly contrite 'Yes miss' and its humorous extension by Paul in schoolteacher role (l.10) are clearly designed to membership Kate as a schoolteacher, exploiting the standardised relational pair, teacher–student (Sacks, 1972; for general introduction to membership categorisation, see Hester and Eglin, 1997). Since this also moves her from a Member of the category 'expert', it undermines the force of her suggestion.

In fact, as so often with humorous interpolations, there are undercurrents to the exchange that are not clear from its surface form but discernible from surrounding features. Jenny's emphatic and repeated 'Mmm' follows Kate's mention of liaising (l.1) which she picks up in her criticism of Kate (ll.14–16). This sensitive issue concerns the distribution of students in joint projects involving the two schools in the organisation, which is in turn part of a wider concern about relationships between the partners. Jenny's charge that Kate has forgotten to send students to the Pen memberships the latter as the head of an organisation who is failing in her responsibilities, and in a passage of only 17 lines Kate has now occupied three Member categories: expert advisor, bossy teacher and deficient head of organisation. In the process of this her position has shifted from that of invited participant to unhelpful outsider, the transition negotiated via the aggressive thrust of the group's humour.

The benefits of humour directed at a third party in terms of generating rapport has been recognised (e.g. Norrick, 2003) as has its use by superiors to maintain their position and subordinates to challenge those above them in the hierarchy (e.g. Holmes, 1998, 2000b). However, authority need not be external; it may be associated with the role someone is fulfilling. There is one example in the Pen data that suggests that Jenny is quite prepared to exploit its potential for humorous reversal. In Extract 7.4, Paul is taking the minutes as the team are covering a list of points they wish to make with regard to possible appointments. Annette has just moved to 'the next one':

Extract 7.4

```
001  Annette:  Salary in line with ⌐Pen Inkham. ⌐
002     Paul:                      └Hold on. No.┘Hold on.
003            Hold on. So:: (.) e:m
004            (0.5)
005     Paul:  ((Reads aloud as he writes minutes.)) 'EC'
006            (0.5
007     Paul:  (things)
008            (1.0)
009     Paul:  becoming more and more necessary. Is that
010            going to fit in.
011    Jenny:  °Whatever you say Paul.°
012    Harry:  Heheh hh
```

Here, Jenny's mock deference to Paul's power to articulate the group's position gains its ironic force from her status as principal and its humorous appeal from the group's democratic orientation, in the context of which Paul's sudden bureaucratic attention to minutiae seems excessive. As someone who is happy to spend half a morning break on her knees in the staffroom scrubbing at a stain on the carpet while life goes on around her, Jenny is never slow to take a pin to what she sees as inflated behaviour, whether bureaucratic or bossy.

There is also evidence in the literature that humour of this kind may not be uncommon where a colleague has violated collegial norms. Sollitt-Morris, for example, provides the following example, also in the context is a school department meeting in which Zeb is the Head of Department and Chair:

Extract 7.5

```
001  Zeb:  okay let's have a look at this agenda ++
002        examine right we've all got a copy of the third
003        form [laughter] what about the fourth
004  Bet:  no Ann hasn't got one yet
005  Ann:  no I haven't [with fake American accent] mom
006        [laughter]
```
(Sollitt-Morris 1997, quoted in Holmes 2000b: 177)

Her analysis of this demonstrates that it also shares many qualities with the challenge in Extract 7.3, although this time the authority figure is the mother rather than the teacher:

> In this example Bet takes it upon herself to answer for Ann, which Ann clearly does not appreciate. Rather than tell Bet overtly that she does not want Bet to speak for her, Ann agrees with Bet then calls her 'mom' in a silly voice. 'Moms' speak for children, and 'moms' have a higher status in relation to their children. By addressing Bet as 'mom', which she is patently not, Ann is able to undermine Bet's interference.
> (Sollitt-Morris 1997, quoted in Holmes 2000b: 177–9)

These exchanges, and the humour associated with them, depend on membershipping the recipient in a way that is inappropriate to the professional circumstances of the encounter. However, the humour directed at the insiders, Bet and Paul, is accomplished in a single turn and, unlike Kate's experience, does not involve other members of the group or lead to a more direct challenge on a related issue. While all such humour is potentially damaging, extended baiting of the victim seems to be reserved for the outsider.

There is no suggestion of humour in an encounter from the IEC, where Tony challenges an outsider in a way very different from that associated with in-group disagreements. The setting is a marking moderation meeting involving two outsiders invited to attend because they will be participating in the assessment of assignments in one of the modules on

the Masters programme. Tony has awarded a different grade from that
of Thomas, one of the outsiders, and challenges the latter on the basis
of Thomas's application of the relevant marking criteria:

Extract 7.6

```
001   Tony:   Because I wasn't struck here (.) by the same
002           things you are.=I wondered if you'd got any
003           (.) particular instances where you could say
004           'Look,
005           (0.4)
006   Tony:   here's this person
007           (1.0)
008   Tony:   referring to this idea (.) and not evaluating
009           it sufficiently or referring to these
010           authors ⌈wi  ⌉ thout
011   Thomas:        ⌊Not-⌋
012           (0.2)
013   Tony:   ⌈⌈o:nes
014   Thomas: ⌊⌊Not ev-
015           (0.2)
016   Thomas: Not evaluating it in ⌈te:rms   ⌉ of (.) the:=
017   Tony:                        ⌊relevant ⌋
018   Thomas: =learners he's talking about. And in terms
019           of the: educational the socio-cultural
020           background from which they come, and not
021           evaluating
022           (0.8)
023   Thomas: what is sai:d about the topi- >what has been
024           said< about the topic in terms of what he's
025           doing with it.
026           (1.0)
027   Thomas: In the assignment.
028           (1.0)
029   Thomas: That was my second po ⌈int.
030   Tony:                         ⌊Er- I hear a sliding off
031           to context but that's e-em-=
032   Jill:   =Yea:h, I's gonna ⌈sa:y⌉ I think-=
033   Jack:                     ⌊I e-⌋
034   Jack:   =OohGod-Hehhehe ⌈hehe but this is one⌉ of the=
035   Jill:                   ⌊Heh heh I (was xxxx)⌋
036   Jack:   =problems ⌈I have⌉ is that it's=
037   Tony?:            ⌊Yeah ⌋
```

```
038     Jack:   =like
039             ┌(someone else is xxxxxxxxxx) YE┐ he ┌hah.┐
040     Jill:   │ Maybe that's there-┐ that's the┘    │    │
041     Tony:   └(xxxxxxxxxxx again)┘                  └(Can ┘we)
042     Tony:   move on to that (.) specific point.
043             ┌Are┐ there instances here you think (.) whe:re
044     Thomas: └Mm ┘
045             (0.4)
046     Tony:   'Here, this guy's: referred to: : Esterhazy
047             eighty-six:, and
048             (0.8)
049     Tony:   and Smith ninety-four, and Jones:=eighty(.)
050             two::=Why?=What for.=
051     Thomas: =Not why- not why what fo:r,
052             (0.4)
053     Thomas: m ┌ore APArt >from for me to ┐ say< (.) did-=
054     Tony:     └Okay so you did (xxx that)┘
055     Thomas: =did- (.) is there evidence that this
056             person has
057             (0.4)
058     Thomas: covered the reading. Y:eah.
```

Before analysing this extract in more detail, it may be useful to summarise the way it develops (the line references are to the main elements and do not take account of overlapping):

01–10   Tony signals disagreement with Thomas's comments and challenges Thomas to produce evidence from the student script in support of his view that the student has referred to the literature and ideas in the literature without critically evaluating them.

11–29   Thomas does not respond directly to this challenge but claims that the student has referred to the literature and ideas in the literature without relating them to the learners and their background or to the argument of his assignment.

30–31   Tony rejects Thomas's move by categorising his response as more appropriate to 'context' a different marking criterion from the one they are currently discussing and one not relevant to the issue he has identified (the student's reference to the literature as part of a developing argument), which falls under the criterion 'argument'.

32–40   Jill and Jack shift the topic away from the one that has become contentious and towards the broader issue, raised earlier, of deciding which criteria apply to which comments.

41–50   Tony recycles his earlier demand for evidence in the form of a hypothetical example: references from the script the relevance of which is not made clear.

51–58   Thomas explicitly rejects Tony's formulation and categorises the failing as the absence of any evidence other than that the student has actually covered the reading.

The strategies used by the interactants here are interesting. Thomas, for example, avoids the specific examples requested by Tony and offers general characterisations of his position. His first response picks up Tony's 'not evaluating it' but redirects it. Instead of providing an example where an idea is not evaluated or supported by references to the literature, he focuses on the areas that he sees as relevant to such an evaluation: the learners and the socio-cultural context. In his second response he explicitly rejects Tony's direct questions and provides an even more general response than his first, in fact a reply that is vague almost to the point of incomprehensibility.

In responding like this Thomas is pursuing a strategy not dissimilar to the one that IEC colleagues employ in arguments, as Chapter 3 shows. Tony's approach, however, is very different. He begins with a statement that the two see something very differently, but in contrast to the earlier encounter, where he positioned himself as someone who had failed to see what the rest of the group had seen and then went on to seek a clarification of this, here he moves straight from a statement that his position is different from that of Thomas to a request that Thomas provide examples to support his own position. Despite the subtle hedging in the way the request is formulated ('I wondered if...'), this is nevertheless a direct challenge to Thomas.

Tony's response to Thomas's answer (ll.30–1), cut off by Jill and Jack's side sequence, is immediately evaluative, assessing the response as one that properly belongs under another criterion (context) and not the one currently being discussed (information and argument). Again this is slightly hedged, this time by the suggestion that Tony 'hears' it this way, but when he pursues the point after the interruption (ll.41–51) his challenge is more direct, calling for specific examples in support of the response Thomas has just made and concluding with two blunt questions: Why? What for. How different this is from his contribution to the group argument described in Chapter 3 where, even when he is presenting a point in direct opposition to one that has already been made, he formulates it as a personal view, even when he could easily demand evidence in support of a new claim that Jack has just made, as in, 'But what you're introducing <u>now</u> as far as I'm concerned is something that- is <u>to</u>tally <u>ne:w</u>'.

No such consideration is extended to Thomas; neither is the relationship one of equals exploring their different positions on an issue. In this case Tony represents himself as the assessor of Thomas's efforts. By asking for supporting evidence from Thomas, Tony establishes the latter's position as the one under challenge, while Thomas's strongest line of response is merely to reject the linguistic forms used by Tony ('Not why- not why what f<u>o:r</u>'). What follows this, overlapping Thomas's continuation, is the immediate reassertion of Tony's role of evaluator 'Okay so you did (xxx that)' (l.54).

This situation is, of course, unusual in that Tony is an established marker and Thomas a newcomer to the game – it would be strange indeed if the roles had been reversed. Nevertheless, the difference here arises not just from the knowledge differential and its consequences but from the fact that Thomas is not an insider, so the interactional routines and ways of speaking established by and for the group do not extend to him.

Invoking the other

Identity is a resource that can be strategically deployed in order to achieve personal, institutional and social goals, sometimes at the expense of relationships which might otherwise be taken as given. The 'other' that is invoked exists as such not by virtue of its imperishable externality but because this is the way it is constructed in the moment, its position ratified by participants in the talk. Once invoked as an external element it exists outside the boundaries of the group and its associated protections and powers, a legitimate target for exploitation. An incident in a Pen staff meeting illustrates how a subtle shift of identity, in this case involving the Pen itself, can serve strategic ends.

When the Pen teachers have in mind 'the school' or 'the Pen', they think of their own school, including its culture and the management structure that exists within it. The teachers recognise that higher level decisions are made at the parent school in Inkham, but discussions of what happens 'there' rarely feature in the data. There is just one example, though, where someone in the Pen core staff distances himself from the school in order to invoke it as a relevant 'other'. In a staff meeting where Jenny is absent, Ed has asked if he can 'butt in' to introduce 'something completely different' to do with his dissatisfaction with the post of social organiser, and when Paul has suggested delaying the item until 'AOB' has rejected this. Ed's complaints are nevertheless heard sympathetically until he suggests that he should have control of the petty cash, which is Pat's preserve:

Extract 7.7

```
001           Ed:  I think the person who's running the
002                social programme should be doing all of it.
003                And (em)
004                (0.6)
005           Ed:  that means access >I assume< to petty cash,
006                that means (.) ┌doing┐ everything.=
007  Annette:                    └Mmm ┘
008      Paul:  =Wo:::::: you've
009             ┌no┐ chance ┌in ┐this place.┐
010  Annette:  └No┘        │Hah│           │ =
011    Harry:              └HEH┘ Eheheheh  ┘
012  Annette:  =°Heheheheh°
013           Ed:  E:m:
014                (0.5)
015           Ed:  because basically it involves too many
016                people.
```

There are only two references to 'this place' in the Pen data, both occurring in this discussion, where it distances Paul – and, since they join in with laughter, Annette and Harry – from the school. The expression might even be taken as mildly pejorative, which serves only to strengthen its strategic purpose: to establish that further debate along these lines is pointless, since nobody at the meeting is in a position to alter the ways in which things are done in 'this place'.

This strategy becomes more explicit as the discussion develops, Paul's invocation of the other responding directly to Ed's moves to reintroduce the cash issue. There is considerable sympathy for the difficulties associated with the job of social organiser and its lack of reward, and at one point Paul reverts to 'we' to represent the school. However, when Ed refers again to access to cash, Paul invokes 'the school' as a separate entity: 'Well this is- this is betwee:n whoever (.) wants to be social organiser (.) and the school to sort out.' The discussion shifts again to the low profile of the job, the 'hassle' involved and the poor pay, reaching a point where the topic can be closed following supportive statements from the core staff. However, the apparent rapprochement is shattered when Ed returns to 'getting your hands on the old purse strings', prompting firmer closure from Paul:

Extract 7.8

```
001      Paul:  Well it's always been a tradition in this
002             place that money is only dealt with by two
003             well basically one.
004        Ed:  °Yeah. Mm°
005      Paul:  And that's (above) two people.
006        Ed:  °Mmm°
007   Annette:  °Yeah.°
008        Ed:  (xxxxxxx)
009      Paul:  I think it's something you're going to have
010             to sort out with senior management.
011             (1.5)
012        Ed:  I mean when it's on school (.) business
013      Paul:  °Uhuh°
014             (5.0)
```

The appeal to 'tradition in this place' fixes current arrangements in an external temporal as well as institutional context, a move that is neatly echoed in Paul's later turn (ll.9–10), where the resolution is shifted to a later time and characterised as an engagement between Ed and a non-present group who represent 'this place'.

What is particularly interesting about this sequence is the way in which the school, an institution created by the core staff with which they intimately identify, has been invoked as an external agency over which they effectively have no control. The fact that this flies in the face of the care they normally take to distinguish their school from the senior management located in the parent school merely underlines the importance of not taking identity for granted. If such a subtle shift of identity allows Paul to position himself in a way that closes the argument down, this is an interactional move he is perfectly prepared to make.

In the IEC argument about the European working time directive, Jack invokes an external agency in support of his position, although this time the other is one with which all his colleagues are well acquainted:

Extract 7.9

```
001    Jack:  I just er:: (.) I don't know (xxxxxxxxxxx
002           when you) have records.
003           (0.5)
```

```
004   Jack:   Yeah they're not gonna be looked at, but
005           what if there's an issue,= what if someone
006           in the business school,
007           (0.4)
008   Jack:   what if someone down in
009           (0.2)
010   James:  ((Coughs))
011   Jack:   in the centre down there's decided they
012           wanna have a GOOD look at the IEC,
013           (0.2)
014   Jack:  ┌┌They're┐ on paper.
015   Paul:  └└Yea:h. ┘
016           (0.2)
017   Kathy:  Ex┌actly
018   Jack:     └They're there.
019           (0.2)
020   Kathy:  °Yeah°=
021   Jack:   =You know.I- this is I- (.) that's what-
022           (.) bothers me a (.) the division in that.
023           In PRINCiple I agree >with the idea of<
024           forty-eight hou:rs: an' I think th't (.)
025           maybe it would be a good idea if we all
026           informally agreed that we would DO THAT.
027           That we w'ld- (.) as you say, quickly
028           reflect at >the end of the day< how many
029           hours we've ┌spent┐ that day,=
030   Tony:              └Yeah ┘
031   Jack:   =and get s:ome sort of picture >'f how many
032           hours< we're doin', but what I object to (.)
033           is giving them that in the form that they
034           want it.= An' giving them any ammunition to
035           use against us.
```

Tony has asked that the group consider completing the relevant forms showing their working hours, and here Jack tries to establish a position where he agrees with Tony's general argument while resisting the precise action he advocates. He does this by suggesting that a written record would leave the group exposed to potential threats from elsewhere in the university.

The threat is first of all built up in a three-part structure based on 'what if' (ll.5–8), culminating in a reference to the centre 'down there' (with 'down' repeated) where someone might want a '<u>GOOD</u> look at the IEC'. This is followed by an emphatic reminder (ll.14 and 18) of the consequences of Tony's proposal: 'they're on p<u>a</u>per ... they're <u>there</u>.' Jack then articulates his position in a way that accepts the principle of Tony's proposal while rejecting the written record, which would serve as 'ammunition to use against us.' An apparently innocent administrative action is represented as a potential weapon: the centre is implicitly the enemy, a threat to the group's well-being.

What both of these examples demonstrate is how the professional identity of the group as a distinct entity allows it to be positioned vis-à-vis an external agency, either real or created, for the purposes of achieving immediate interactional ends. It is this, I would like to suggest, that distinguishes *invoking* the other from *representing* the other. In the former case, the other is treated merely as a device that allows the user to achieve interactional ends, its use being at least in part dependent on the group's shared understanding, or at least acceptance, of the relevant characteristic(s). In the latter case, however, the other needs to be more carefully constructed as a relevant, if non-present, participant in the ongoing talk, as the next section will demonstrate.

Representing the other

Linguistically, interesting things happen when a non-present other is introduced into the talk of a collaborative team. This section offers an example from each setting indicating the range of devices exploited in the relational positioning that such events occasion. The first, and perhaps most straightforward, example again reflects John's slightly ambiguous position as head of DOTS, uneasily poised between team member and participant in higher level decision-making, a situation that calls for some delicate manoeuvring when representing the position of higher management.

Dealing again with the touchy issue of pay and grades, John has just explained that for research scientists and above there will have to be consultation outside the team about the standards that will apply within pre-determined categories. Julie, though, is sceptical about the extent to which such consultation will be genuine:

Extract 7.10

```
001   Julie:  But when they say consultation they
002           don't wan- so much mean consultation as
003           dictation.
```

```
004   John:    ⌈⌈No::,  ⌉ at the very be-
005   ????     ⌊⌊ Mmm   ⌋
006            ((Suppressed laughter from a few of those
007            present.))
008   John:    At the beginning it's a consulta:tion
009            (2.5)
010   John:    amongst °er° CERtain people, to decide what
011            the standard should be .=
012   Julie:   =Mmm
013            (1.0)
014   John:    Subsequently there will be consultation
015            (1.5)
016   John:    to ensure that
017            (0.2)
018   John:    Penny being an 'SSO' is treated (.) equally
019            fairly to an 'SSO::', in another centre.
```

Julie's representation of what 'they' have in mind, and the suppressed laughter which follows, leaves John in the awkward position of having to represent 'them' without explicitly aligning with them. His solution is simple: remove the agent. What follows, carefully elaborated with long pauses between points, manages to represent the situation without placing those responsible either inside or outside the group. First he finesses the issue of identification by saying that the first phase of the consultation will involve 'certain people' who will decide on the relevant standard. The second phase, indicated by 'Subsequently' in line 14, also involves consultation and again the identity of those involved remains undisclosed. It is a simple strategy but by leaving the other unspecified John has avoided the possibility that he might have to associate himself with an outside entity.

The relevance of this strategy of evasion has been noted by Diamond:

> Interactions in which interlocutors use evasive and politic language indicate a situation which is threatening due to some background information, the constellation of participants, or a future intention, not only a particular act or request that needs redressive action.
>
> (Diamond, 1996: 63)

It applies equally well to the next situation, where again there is no question of the need for redressive action. In this case the speaker, again the head of the group, uses politic rather than evasive language to achieve the necessary outcome. It takes place in a meeting at which a representative of the library, Gail, has been invited to discuss a joint project, one that would normally involve the School's IT officer, Ben. However, the team are concerned that he should not be involved in the early stages because he is already severely overworked and there is a need to ensure that he is able to concentrate on other ongoing projects. As Gail knows, this is really a decision for Ben himself, so the matter, as Jack notes, is a delicate one: 'It's de- delicate isn't it. Of course. (0.4) °Delicate.° (1.0) E::m (0.5) it's all °delicate here. °' His opportunity comes when Gail assents to James's suggestion that the matter be treated as an information issue rather than a technology issue:

Extract 7.11

```
001  James:   Em I think if this could be (.) treated as an
002           information issue
003  Jack:    Yeah=
004  James:   =rather than a technology issue, I'd
005  Gail:    That's fine.
006  Jack:    So that's ┌a (xxxxxxxxxxxxxxxxxxxxxxxxxxx)┐
007  Mark:          │    ((Clears throat))             │
008  Gail:         └I just want to know when  (xxxx) ┘
009           this because I don't want to
010           (0.3)
011  Gail:    em- you know I don't want to upset him
012           Basic┌ally.
013  Jack:         └OH NO I wouldn't want to upset Ben but
014           I-I feel that the problem is that he would
015           feel perhaps professionally obli:ged ((More
016           deliberate articulation of consonants
017           resumed)) to be involved when it's clear that
018           he's already in a situation where he's not in
019           a position to provide the services that he's
020           expected to provide. And I think it would be
021           unfair (.) to place that additional
022           burden ┌on him┐ of asking him
023  Gail:          └mmm ┘
024           (1.5)
```

What is most striking about Jack's response in lines 13–22, apart from the unusual deliberateness of its delivery, is its length (66 words) and complexity. It is of a piece with other examples in the data where this mode of delivery marks a statement that Jack intends to be used as an official representation of the group's position. Typically, such statements, which may subsequently be repeated and noted in shorthand by Sally, do not represent the group's true position but their official line; hence they reflect the group as a responsible other within the organisation. In this case the representation has embedded within it an interpretation of the outsider Ben's position as someone who would feel obliged to accept the 'burden' of an invitation to participate. The group know that Ben would not see this as a burden at all and would be eager to participate, but the mode of delivery indicates that an objection along these lines would be inappropriate.

The subtleties of this, however, are as nothing when set against the way the Pen group deal with a potential criticism from one of their students. Ed introduces this into one of their meetings, a rare example of student dissatisfaction arising from an end of course evaluation:

Extract 7.12a

```
001     Ed:  Does Alan know that- you know that (.) you
002          know (like) Anne Grest (.) did they know that
003          >they were kind of< coming (.) to a general
004          English type course.
005  Jenny:  °Uhuh°
006     Ed:  Uhuh
007  Jenny:  As far as I know. That's the information they
008          should have been given.
009          (1.0)
010  Jenny:  Why: (.) did you get the
011          impression ⌈the other te-
012     Ed:             ⌊Well Nigel seemed to give the
013          impression that he (.) wasn't (.) he didn't
014          get quite what he was expecting.
015  Jenny:  Did he?
016     Ed:  Yeah.
017          (1.0)
018  Jenny:  O::h. This could be a fault at the British
019          Council end I suppose. What did he think (.)
020          he was going to get.
```

Ed introduces the criticism in the form of a general question (ll.1–4) which serves as a pre-sequence (for a discussion, see Levinson, 1983: 345), allowing Jenny to direct responsibility away from the school (ll.7–8), by means of what might be called *pre-emptive positioning*, before inviting detail of the criticism itself (ll.10–11). The process by which the complaint is actually introduced is an interesting one, especially in terms of the control Jenny exercises over it. Once the initial question and answer pair is complete the talk proceeds in the form of question and answer adjacency pairs controlled by Jenny, who asks the questions (ll.10, 15, 19). The hedged formulation of the criticism ('seemed to give the impression...didn't get quite...') is almost invited by the question which prompts it: 'did you get the impression...'. There then follows a challenging 'Did he?' (l.15) and an attributable silence after Ed's response. Jenny's response, when it comes in line 18, is preceded by a change of state token indicating surprise, something that will be relevant later when inconsistency on the part of Nigel is indicated. Jenny's responses so far have been a mixture of challenge and deflection, and her explanation (ll.18–19) sums this up, first picking up the earlier pre-empt in referring to the British Council and then suggesting a misconception in the emphasis on 'think' as she directs another question back at Ed.

At the end of this remarkable sequence, Jenny has succeeded in assigning responsibility for a problem before the problem has actually been identified, then inviting a formulation of the problem that will diminish its force. In comparison with this wonderfully economical piece of interactional positioning, what follows is almost clumsy. It begins with Ed's response to Jenny's question, 'What did he think (.) he was going to get':

Extract 7.12b

```
021    Ed:    Well I mean he- (.) he just said on- on the
022           Friday >he said< 'I don't- I don't kno:w (.)
023           how much I (.) improved,' >he said< 'partly
024           because I'm (.) slightly out of my depth,'
025           (.) he's-
026  Jenny:   Mm
027    Ed:    'and- and ┌partly because      ┐
028  Jenny:             └(I understand th┘at.)
029    Ed:    because some of the:' (.) >you know<
030           'materials w-'
031           (0.6)
032    Ed:    some of the things about (xxxxxxx) to him.
033           Which was fair enough
034           (0.6)
```

```
035      Ed:  I suppose but
036           I ⌈mean⌉ (.) anyway (.) the kind of=
037  Jenny:     ⌊Mmm⌋
038      Ed:  =course they're coming on
039           (0.6)
040  Jenny:  It's probably worth a (.) a note back to (.)
041          the British Council to make sure this gets
042          conveyed.
043      Ed:  Mmm
044  Jenny:  Er (.) I don't know how well liaison goes on
045          at that end,
046          (0.6)
047  Jenny:  I know some of the (.) mistakes they've
049          made in getting accommodation information
050          through are not that (kind of job but)
051      Ed:  °Mm°
052  Jenny:  But yes it's probably worth finding out to
053          make sure.
054          (0.6)
```

Ed's response to Jenny's question is first delayed by a number of hesitation markers and a hedge ('he just'). The use of 'on the Friday' rather than 'on Friday' at least leaves open the possibility of other statements at other times, the former expression tending to feature in narratives, implying that the incident is one in a sequence ('I visited Tom on the Friday, then we went on to London on the Saturday'). The distancing itself, though, is achieved through the use of direct speech. In his discussion of production formats, Goffman (1981) distinguished between the *animator*, who produces the talk, the *author*, who creates the talk, and the *principal*, who is responsible for the talk. In this case, by using direct speech to represent Nigel's position, Ed establishes Nigel as both author and principal, with Ed as merely the animator. Indirect speech would have transferred authorship to Ed, but by maintaining a minimal role he is able to distance himself not only from the complaint (for which the principal is responsible), but also from the complainant (who authors the complaint).

What follows involves an interesting shift of footing (Goffman, 1981) from the evaluation of Nigel's claim, 'Which was fair enough', as a teacher on the course and a member of the group to which the comment has been presented. Following a brief pause during which nobody offers a response to this (1.34), he hedges it ('I suppose') and begins a qualification ('but') to which Jenny responds ('Mmm'). Although Ed's position on 'the kind of course' is never clearly articulated, the 'but' (l.35) aligns him with those who question the validity of the criticism. From this point on, Ed's primary interactional endeavour is that of alignment with his colleagues and therefore against Nigel. This grows in strength, from receipt tokens (ll.43 and 51), through explicit agreement (ll.55 and 77), to jointly constructed evaluation (ll.76–90).

Jenny meanwhile suggests an appropriate response from the Pen and then contextualises this misunderstanding in terms of failures 'at that end' which are represented as numerous (1.47: 'some of the mistakes'). Her suggestion in line 52 that it is 'probably' worth finding out implies that such an action is not essential because the cause of the problem is clear. This effectively deals with the problem at the British Council 'end', but the dismissal of the criticism does not end here:

Extract 7.12c

```
055   Jenny:  But apparently he went with great (.) thanks
056           and
057     Ed:   Yeah.
058   Jenny:  praise and everything.
059    Paul:  Who
060   Jenny:  Despite (.) Nigel
061    Paul:  Nigel.┌Yeah.┐Considering he=
062   Jenny:        └Mmm ┘
063    Paul:  =went (.) extremely lazy
064           (0.6)
065   Jenny:  °Mmm°
066    Paul:  Er I- I mean i- it could have been: (.) real
067           problems about loss of face, it could have
068           been that.
069   Jenny:  Mmm. I felt
070    Paul:  E:m
071   Jenny:  he was worried about his status ┌(xxxxxxx)
072    Paul:                                  └But he- he:
073           he he- he didn't work well within the
074           group ┌after-┐ you'd bear me out on this.
075   Jenny:        └Mmm  ┘
076    Paul:  He ┌he was ┐ I felt sorry for anybody working=
077     Ed:      └Yeah he┘
```

```
078   Paul:   =with ┌him
079   Jenny:        └Mm
080     Ed:   I think he felt a bit like
081           ((Tape snarled))
082     Ed:   I mean he's the president of the Chamber of
083           Commerce °hh heh ·hh hh° and ┌er ┐ George was=
084   Jenny:                                └Mm ┘
085     Ed:   =the director of the Chamber of Commerce, and
086           I just think generally ┌that ┐
087   Paul:                          └Geor┘ge (.) seemed
088           very appreciative.
089     Ed:   Mmm=
090   Jenny:  =Mmm=
091   Paul:   =(Some.) Again this is (xxxxxxx)┌to reason.
092   Jenny:                                   └Mmm
093   Jenny:  Yes George ┌was┐ incredibly nice to me=
094   Paul:              └Er ┘
095   Jenny:  =when I (xxxxxxx) farewell to
096           him. ┌(xxxx) with him he was very ┐ good.
097   Paul:        └And I had a chat with him.   ┘
098   Jenny:  (Gallant), °and°
099   Paul:   And he ┌said┐ he he (.) he's really got a lot=
100   Jenny:         └(xx)┘
101   Paul:   =from the course.=
102   Jenny:  =Yes.
103   Paul:   Because I asked him and ┌he's-┐ (.) he was=
104   Jenny:                           └Mm   ┘
105   Paul:   =very (.) very (.) clear about it,=
106   Jenny:  =Mm=
107   Paul:   =and >kind of< (.) >you know< (.)┌sort of
108   Jenny:                                    └Mm
109   Paul:   =(just) (.) just (xxxx) by a stra:nge
110           person.=
111   Jenny:  =Mmm. I think yes they (xxxxxx).
112   Paul:   It ┌was  ┐ strange rather than: sort of=
113   Jenny:      └(xxx)┘
114   Paul:   =em (.) aggressive I felt.
```

Attention now shifts from the complaint to the complaining student and is designed to undermine the legitimacy of his position. Jenny's immediate observation (which gains Ed's assent) that he left having thanked and praised them (ll.55–8) implies at best inconsistency and at worst duplicity. Paul joins in to suggest that Nigel 'went (.) ext<u>re</u>mely lazy', thus assigning the cause for his failure to improve to a failure in his own character. Paul and Jenny's tentative suggestions that his loss of face and worry about status might also have been a factor are followed by Paul's more confident assertion that he did not work well within the group and an appeal to his colleagues to support this contention. Paul then expresses sympathy with other members of Nigel's group, thus subtly aligning the teachers' position with that of the rest of the students. All that remains is to call on a specific example, George, as representative of the rest of Nigel's group. George's position is represented unequivocally: he 'seemed very appr<u>e</u>ciative' (Paul, ll.87–8), 'was incr<u>e</u>dibly nice to me' (Jenny, l.93), and was 'very (.) <u>clear</u>' that he 're<u>all</u>y got a lot from the c<u>our</u>se' (Paul, ll.99–105). A final observation, that Nigel was a 'strange' person, confirms his deviancy and draws spurious professional legitimacy by its placement within a superficially balanced assessment recognising his lack of aggression.

A summary of the moves in this exchange demonstrates just how effectively this challenge to the professional competence of the Pen staff has been dismissed:

Presentation A pre-sequence allows pre-emptive positioning before the criticism itself is formally requested and presented. By presenting the criticism in direct speech, the speaker effectively distances himself from it and sets up his subsequent alignment with colleagues as they respond to the criticism.

Deflection The pre-emptive positioning is taken up, as responsibility is explicitly shifted to an outside party.

Typification The complainant is typified as a student who has problems or defects that set him apart from other students.

Isolation The complainant is presented as someone who does not fit in with the rest of the group. Hence, implicitly, his views are dismissable as deviant.

Contrast The contrast is made explicit by taking another group member and positioning him at the opposite extreme to the complainant.

By the end of this sequence the force of the complaint has been directed elsewhere and its validity undermined. The Pen staff have located the problem giving rise to the complaint outside the Pen, in the British Council, and have effectively identified the source of the complaint itself, Nigel, as an atypical student. The group has preserved its professional face, but at the expense of an external organisation represented as systemically inefficient and a student portrayed as unreliable. The consequences of this and the other examples in this chapter, it could be argued, serve the group's immediate ends, but the long-term implications might be less beneficial, as the final chapter will suggest.

Facework

Pervasive though facework is in our everyday encounters, I have delayed its consideration until this point because of the difficulties it presents when used as an analytical resource. My position here follows that of McMartin et al. (2001) who argue that although the concepts of face and facework are of fundamental importance, many of the criticisms that have been directed at them derive from their use as a priori categories in analysis. The authors argue for the conversation analytic position, which rejects the use of concepts such as face in the process of analysis but is nevertheless prepared to deploy them when interpreting and discussing analytic findings. I therefore use the examples in this chapter as a convenient point of departure for a brief consideration of face issues in the interaction of collaborative groups.

The concept of face was first introduced by Goffman, who defined it as 'the positive social value a person effectively claims for himself' (1967: 5), but it was Brown and Levinson (1978/1987), drawing on Goffman and on Grice's (1975) work on conversational implicature, who developed the fullest account of its nature. Crucially, they drew a distinction between positive face, the desire for solidarity with others, and negative face, the desire that one's actions should be unimpeded by others. Politeness strategies, they argued, are designed to minimise face-threatening acts (FTAs). Their analysis also proposes that the use of such strategies is influenced by considerations of the degree of imposition involved, the power relationship that obtains and the degree of social distance between speaker and hearer.

This very brief summary does scant justice to the complexities of the authors' analysis (for an excellent brief summary, see Diamond, 1996: 48-83), but at its heart is a recognition of the fundamental importance of preserving an individual's self-esteem and the interactional implications of this. One way of approaching the issue of face in this context without resorting to speculation about mental states is to examine the sequential development of relevant exchanges from the point of view of preference organisation. Although the concept itself is not without its own analytical dark side (see Boyle, 2000), its focus on turn design throws interesting light on conversational facework as a brief analysis of extracts from the exchanges in this chapter will reveal.

Preference organisation recognises that sometimes speakers design their turns in such a way as to elicit a particular response, so that for example a question would normally be designed to elicit an answer and an invitation an acceptance. When the expected response is forth-coming, this is described as preferred, when it is not, it is dispreferred. Consider, then, Kate's injunction in the following:

Extract 7.3a

```
004   Kate:   Well just-=
005   Jenny:  =Mmm=
006   Kate:   =think about it will you.
007   Paul:   hh hh ⌈hh hh heh
008   Jenny:        ⌊heh heh heh ((Suppressed laughter.))
009   Jenny:  Yes miss.=
010   Paul:   Come on. >Which one of you< forgot that.
```

Her turn is designed to generate assent, but instead Paul laughs and Jenny tries not to. The assent, when it does come, is in the form of a child's response to a teacher and a follow-up to Kate's request made by Paul who, as a recipient, is not sanctioned to do this. Sequentially, then, there is ample evidence that the normal response pattern has been subverted. The fact that Kate's injunction is a bald, on-the-record FTA without redressive action, threatening the group's negative (and arguably positive) face (Brown and Levinson 1978/1987: 65-70), accounts for what follows in the sequence, itself an FTA directed at Kate's positive face. Additional mentalist accounts of what may or may not be driving these acts are superfluous.

Suppressed laughter again features in the following extract, where John's turn in response to Julie's claim is interrupted:

Extract 7.10a

```
001   Julie:  But when they say consultation they
002           don't wan- so much mean consultation as
003           dictation.
004   John:   ⌈⌈No::,⌉ at the very be-
005   ????    ⌊⌊Mmm ⌋
006           ((Suppressed laughter from a few of those
007           present.))
008   John:   At the beginning it's a consulta:tion
```

Here, the laughter is not timed to occur at a transition relevance place but instead interrupts John's explanation in line 4. Whether or not the laughter is directed at him is problematic because, while its placement might indicate this, Julie's challenge to authority seems a more likely candidate. However, this is irrelevant to an analysis in terms of the turn-taking system: John has the floor, has just begun a dispreferred response to Julie's claim and is interrupted. The fact that he does not join in the laughter and recycles his turn signals a lack of alignment with the laughers, treating the laughter as disruptive.

It would be possible to analyse every turn in this sequence, with the exception of line 5, as potentially face-threatening, but such an unwieldy approach would add nothing to the essential point that the disrupted sequence represents a threat to the group's unity. The same consideration applies to the concepts of institutional face (Tracey and Carjuzá, 1993: 188) and even group face (Nwoye, 1992: 313; Spencer-Oatey, 2002), which in any case do not seem to offer sufficient analytical purchase to make their adoption worthwhile. However, Holmes's more general claim (2000a: 48-9) that '"Doing collegiality" involves, in particular, paying attention to the positive face needs of participants' seems to capture very succinctly the broader context within which such exchanges should be seen.

Where a participant is not a full member of the group or an outsider, 'doing collegiality' is no longer an issue and the need to address positive face drops away, as Extracts 7.3 and 7.6 illustrate. The same applies to Ed, who is a part-time teacher pursuing an issue unsupported by the core team:

Extract 7.7a

```
005        Ed: that means access >I assume< to petty cash,
006            that means (.)┌doing┐ everything.=
007 Annette:              └Mmm ┘
008      Paul: =Wo::::::you've┌no┐ chance┌in ┐this place.┐
009 Annette:                └No┘        │Hah│            │=
010     Harry:                          └HEH┘Eheheheh    ┘
011 Annette: =°Heheheheh°
012        Ed: E:m:
013            (0.6)
014        Ed: because basically it involves too many
015            people.
```

This time the normal rules for turn-taking are not disrupted but Paul's turn, which latches on to the end of Ed's, provokes laughter that interrupts the normal flow of talk. Since Ed has just placed an assumption before the group, his turn is designed to elicit a positive assessment of that. What emerges is not an assessment, but a dismissal followed by laughter, both face-threatening. When the volume of the laughter diminishes, Ed reclaims the floor with a filler (1.12), then provides an extension to his original assumption which provides a justification for it. Like John, he too is a part member of the group, and this may account for his failure to align with colleagues and the pursuit of his original point in a way that avoids direct conflict. This can be contrasted with the response of an outsider:

Extract 7.6a

```
049    Tony:   and Smith ninety-four, and Jones:=eighty(.)
050            two::=Why?=What for.=
051    Thomas: =Not why- not why what fo:r,
052            (0.4)
```

Here there is no disruption to the turn-taking but a clearly dispreferred response to a question. Tony has ended his turn with two Wh-questions, the first pair part of an adjacency pair predicting an answer as the second pair part. This is forthcoming but is an explicit rejection of the questions themselves, the silence afterwards leaving the rejection isolated at the end of what might be described in terms of politeness as an exchange of bald on-the-record threats to positive face. How different this is from the delicate interactional work that marks Extract 7.12, where careful positioning isolates and dismisses the threat to face that explicit criticism carries.

Facework, as Goffman and Brown and Levinson showed, is an integral part of all social encounters and this applies as much to these collaborative groups as to any other interactional participants. However, while the negotiation of argument in Chapter 3 demonstrated the groups working to maintain positive face, this chapter has used the analysis of ongoing talk to provide evidence of threats to face and the negotiation of these. The final word on face, though, belongs to the participants themselves. In Extract 7.11 Jack demonstrated in his choice of words that he was very aware of the importance of not threatening Ben's negative face:

OH NO I wouldn't want to up<u>set</u> Ben but I-I feel that the problem is that he would feel per<u>haps</u> professionally obli:ged to be involved when it's <u>clear</u> that he's <u>already</u> in a situation where he's not in a position to provide the services that he's ex<u>pected</u> to provide. And I

think it would be un<u>fair</u> (.) to place that additional burden on him
of <u>ask</u>ing him

What these words do not reveal, however, is what was clear from the
ongoing talk: that these considerations were themselves invoked to
achieve the group's strategic purpose of ensuring that Ben spent his time
pursuing other work.

Conclusion

It has been argued that 'the "culture" of an organization is displayed
through a huge variety of contested "us" and "them" claims' (Parker,
2000: 227), but what this chapter has shown is that the simple opposition this posits is by no means straightforwardly constructed. The apparently clear-cut issue of 'them' versus 'us' as exemplified in Jack's opposition to 'giving them any ammunition to use against us' is not a token
of a particular positional type but a constructed relationship. As Paul's
use of 'this place' showed, the place that is professionally 'us' may be
reconstituted as definitively 'them' in order to serve immediate interactional ends; indeed, the whole issue of the speaker's position in one of
the two categories might be carefully finessed, as illustrated by the lack
of an agent in John's representation of the company's position.

Perhaps paradoxically, the freedom to construct such relationships
is a more convincing reflection of the collaborative group's solidarity
than a monolithic us/them relationship would be. Paul's use of 'this
place', for example, produces no attempted repairs from his colleagues,
presumably because they are able to make the necessary association with
external management. Diamond's claim, then, that the 'continual process
of creating social identities necessitates that verbal interaction is laden
with potential conflict' (1996: 145) needs to be reconsidered in the light of
stable collaborative groups, where there is little evidence of such conflict
within the group but some suggestion of its more pervasive presence in
encounters with outsiders – whether direct or interactively constructed.

However, in constructing the other in ways that suit its immediate
interactional ends, the group may also be isolating itself from outside
influences that could otherwise be beneficial to it. The dangers of
enjoying a joke at the head of the organisation's expense are obvious
enough, but more subtle dangers lurk in interactional routines that cut
off possible criticisms before they are properly articulated or directly
challenge outsiders who might otherwise be supportive of the group.
The implications of this will be considered in the final chapter.

8
Conclusion

> Still, in the end, it is the meanings we produce that matter.
> (Wenger, 1998: 51)

Introduction

It is perfectly natural to think of groups in terms of who they are and what they do, forgetting the linguistic building blocks that give shape and substance to their achievements. Yet what is professionally achieved is also linguistically achieved and the identities that emerge from engagements with daily business are as much linguistic as professional. Although my exploration of this fundamental relationship has established no definitive characterisation of language and professional identity and no key to the mysteries of their relationship, it has perhaps located some interesting coordinates in the world of collaborative groups. In this final chapter I suggest ways in which these might be used to guide interventions designed to improve the work of such groups, but first I return briefly to some important analytical considerations.

Analytical issues

I provided the general case for my analytical approach in the opening chapter, but it is only when the analytical scalpel has done its work that the quality of the surgery can be judged, and the force of some of the points I wish to make depends on this post-operative perspective.

Specifically, I think there is a need to respond to two related questions that cannot be left hanging:

- *How can you justify using such a diversity of analytical approaches?*
- *What has happened to organisations in your analysis?*

The first charge recalls a story that Jefferson tells (2004) of hearing and agreeing with a paper that was then criticised for relying on the use of unexamined Members' categories as analytic categories. Apart from any theoretical reservations about such an approach, the practical flaw in the paper to which she refers arose from its illegitimate assumption that a particular type of behaviour was characteristic of a specific category of persons. This danger is particularly acute when dealing with selective groups such as the three in this study, and it is exacerbated by my decision not to work within the rigorous analytical framework of conversation analysis. A response to it therefore has to address both of these issues.

The more precise point to which Jefferson refers is perhaps only peripherally relevant to the analysis here since I have not used the category '[Member of] collaborative group' as an analytical resource, though I have occasionally brought categories such as 'Principal' or 'Head of Unit' into the interpretive frame and am open to challenge in this respect. The more general point about associating particular behaviours with particular groups is perhaps best seen as a sampling issue, and one that faces any researcher working with case studies. It can be addressed in terms of the wider generalisation debate (see, for example, Gomm et al., 2000; Williams, 2002), but a more succinct response in this case is simply to clarify the status of my claims. I am not arguing that all collaborative groups interact in the ways described here or seeking to escape into a definition of collaborative groups framed in terms of particular interactional characteristics. What I have shown is that a centripetal dynamic is characteristic of interaction in all three groups across a range of talk types, and I do wish to suggest that the evidence here points to the likelihood that this will be found in similar groups. Furthermore, I have produced evidence of associated interactional patterns that could, in particular circumstances, diminish the group's receptivity to external influences. These are modest but I hope not insignificant claims.

The more general issue of analytical eclecticism is part of an ongoing debate that has reached an interesting point in its development. In their introduction to a special issue of *Discourse and Society*, Iedema and Wodak noted that 'investigators are increasingly stepping beyond their original disciplinary boundaries' (1999: 6) and there seems to be a growing acceptance of this as a legitimate stance. What has not changed is the pervasive challenge of establishing convincing links between macro and micro, which brings with it the intractably stubborn problems of context. According to Putnam and Fairhurst (2001) language analysis now occupies a prominent position in organisational studies, but their discussion of eight different analytical traditions highlights how each struggles with different dimensions of the problem of context underlying the problematic relationship between discourse and organisation.

If nothing else, overviews such as this demonstrate the impossibility of resolving contextual issues by recourse to Alexandrian measures: this particular Gordian knot calls for more subtle unpicking. My response, perhaps amounting to no more than a tug at the fibres, has been based on using the fine-grained sequential analysis of CA but drawing on setting specific details where appropriate, an approach sanctioned by other writers in the field (e.g. Iedema and Wodak, 1999; Arminen, 2000) but also recognised as innately challenging (Cicourel, 2003: 371). Its virtues, though, outweigh the accompanying disadvantages, not least because it is not distracted by the pursuit of an impossible ideal, a point captured succinctly by Tracey and Naughton (2000: 79): 'There is no such thing as 20-20 vision for interaction; lenses of one kind or another are required to see.'

Once aspects of context are identified as legitimate analytical resources, issues of breadth become immediately relevant, hence the second challenge: that my treatment has ignored the wider organisations of which these groups are part. An analysis working from the standpoint of organisational communication would depend on pinning down this relationship, but my focus on particular groups calls for a different perspective, one that is not based on a part-whole relationship. Hence my response to the question of what makes practices 'organisational' (Iedema and Wodak, 1999: 10) rests on a holistic definition of organisation as 'a set of people who share many beliefs, values and assumptions that encourage them to make mutually-reinforcing interpretations of their own acts and the acts of others' (Smircich and Stubbart, 1985: 727). An examination of how these beliefs and interpretations are interactively constructed within particular groups reveals how organisation is 'done' as part of everyday business. I have therefore to some extent ignored how institutional practices are managed, so that in Chapter 2 I was less interested in how the decision on funding was reached in terms of organisational rules and systems than in how the group revealed its orientation to unstated norms of behaviour. Similarly, although the (institutional) topic in Chapter 6 was systems and routines, my interest was in how the different groups can be characterised by the way these are interactionally constituted.

The penetration that this perspective allows generates insights into the ways in which things are done but has little if anything to say about broader institutional matters. The remainder of this chapter will illustrate the contribution such an understanding can make, first in so far as it throws new light on problems that have already been identified in the everyday operations of tightly-knit groups, and then from the point of view of what it might offer in terms of staff development.

Grouptalk

The recognition within social psychology that discourse analysis 'seems underutilized as a method for investigating bona fide groups' (Stohl and Putnam, 2003: 411) suggests that there is space within such traditions for insights that work of this sort might offer, but the challenges of hybridity are formidable. All analysts of spoken interaction know the siren call of psychological speculation, and enthusiasm for what might be jointly achieved needs to be tempered by awareness of the risks involved. What follows, then, is a tentative exploration of a single topic in the field of psychology, not with a view to occupation but from the perspective of an outsider seeking to better understand how light from different traditions might illuminate areas that a single beam leaves in shadow.

Janis's choice of the term *groupthink*, with its Orwellian overtones, was a deliberate attempt to suggest the corrosive effects of in-group pressure:

> I use the term 'groupthink' as a quick and easy way to refer to a mode of thinking that people engage in when they are deeply involved in a cohesive in-group, when the members' strivings for unanimity override their motivation to realistically appraise altern-ative courses of action. ...Groupthink refers to a deterioration of mental efficiency, reality testing, and moral judgment that results from in-group pressures.
>
> (Janis, 1972: 9)

A consideration of subsequent research in this area suggests a more subtle picture and one more consonant with the findings of my study than Janis's original stark representation would suggest, although recent events have served only to reinforce the darker side of closed groups. In 2004 a highly critical report by the Senate intelligence committee on the US invasion of Iraq concluded that the US intelligence community suffered from 'collective groupthink' in its assessment of the threat posed by Saddam Hussein and that '[t]his groupthink also extended to our allies and to the United Nations and several other nations as well'. It was perhaps hardly surprising that the Butler Report on aspects of the intelligence procedures of America's closest ally, the UK, drew on the same concept in order to explain shortcomings in the system there.

Although Janis's characterisation of groupthink developed over time (Janis 1972, 1982, 1989), its focus remained firmly fixed on (typically high profile) 'fiascos' such as the US government's failure to anticipate the Japanese attack on Pearl Harbor, its disastrous Bay of Pigs invasion and its attempt to rescue hostages during the Iran crisis. An analysis of these and similar cases, argues Janis, reveals that when a highly cohesive group is under pressure to make a quality decision the pull towards uniformity can be irresistible, closing down options and reducing the chance of a successful outcome. Instead of exploring as many options as possible, the group limits its discussion to a narrow range of alternatives and when it has identified a course of action ignores any new information that might count against this. The result is that members fail to recognise what might go wrong – even disastrously wrong – with their chosen course of action.

The groupthink model proposed by Janis identifies three antecedents: cohesiveness, structural faults in the organisation and a provocative situational context. These antecedents can lead to a concurrence-seeking tendency, the defects of which can combine together to undermine successful decision-making. Groupthink is identifiable in terms of eight key symptoms that provide the best guide to its nature, leading Janis (1989) to insist that 'practically all' of them must be present for it to occur. He divides them into three main types which are then subdivided: overestimation of the group, its power and morality; closed-mindedness; pressures towards uniformity.

Although Janis's theory has had considerable impact, assessments of it have been mixed. Reviewers acknowledge its popularity (Esser, 1998) and intuitive appeal (Aldag and Fuller, 1993) but lament the lack of systematic research supporting its claims (Longley and Pruitt, 1980; Moorhead, 1982; Aldag and Fuller, 1993; Esser, 1998). Fuller and Aldag (1998: 165) are damning in their judgement: 'Despite a quarter of a century virtually devoid of support for the phenomenon, groupthink refuses to die and, indeed continues to thrive.' Their opposition to the concept derives from the extent to which a focus on groupthink has diverted attention away from broader issues of group decision-making, but reviews of research in this area suggest that relatively few studies have focused simply on the phenomenon itself; most have been interested in particular aspects of it.

The most interesting aspect of Janis's theory for this study is his view of cohesiveness, which he sees as the most important antecedent of groupthink and represents, somewhat vaguely, as the 'degree to which members value their membership in the group and want to continue to be affiliated' (1982: 245). Cohesiveness, he claims, 'is a necessary condition' of groupthink. In her study of the relationship between this and leader behaviour, Leana identifies it as 'a property of the on-going groups that emerges and develops as a result of member interaction' (1985: 7), which is what makes the study of such interaction so important. Her work would suggest that the groups in my study, headed by participative rather than directive leaders and with strong group traditions, are not strong candidates for groupthink. However, other researchers (e.g. Tetlock et al., 1992; McCauley, 1989) have questioned whether cohesiveness is in fact a necessary condition for groupthink and challenged the essentially negative cast of Janis's approach. Aldag and Fuller (1993), for example, highlight the importance of group synergies and the benefits of cohesiveness in terms of enhanced communication, decreased tension, etc.

Despite the intuitive appeal of the idea that cohesion enhances performance, a review of over 200 published and unpublished articles, reports and theses by Mullen and Copper revealed 'a mixed and inconclusive stance' (1994: 211). Their survey indicated that a cohesiveness-performance effect does exist to a highly significant degree but is actually of small magnitude. However, research indicates that this effect tends to be greater in smaller groups and that commitment to task seems to be the most important component in terms of its strength. The focus in such research has been on outcomes, often of laboratory experiments, and it has not sought to understand the processes through which cohesion is realised. When these are examined, a complex picture emerges, and one that might account for the rather complicated, even contradictory, outcomes that Mullen and Copper encountered.

To date, none of the work on groupthink and related issues has been based on what real groups actually do in their day-to-day business, or how their decisions are interactively constructed. The body of work currently available has been based on narrative analysis, usually drawing on published reports, but in some cases on less reliable sources – Janis admits that in his analysis of the Bay of Pigs invasion 'the available evidence consists of fragmentary and somewhat biased accounts of the deliberations of the White House group, (1982: 46). This approach has allowed reanalysis using other sources (e.g. Tetlock, 1979), sometimes supported by specific analytical tools (e.g. Tetlock et al., 1992) and has identified features that have been further explored under laboratory conditions. While all this has produced interesting, if sometimes contradictory, outcomes, it is far removed from the myriad quotidian engagements out of which such decisions ultimately emerge. These tiny fragments of talk are the unseen parts of a bigger picture, ultimately contributing to the misleading rationality of *post hoc* representation.

To understand the process of groupthink it is necessary to go behind the narrative representations and laboratory recreations of group decision-making in order to explore the ways in which such outcomes are constructed by and for the group. In this world the 'illusion' of unanimity asserted by Janis hardens into a reality of constructed congruence in which there are genuinely no dissenters. Chapter 3, for example, showed how much interactional work goes into avoiding argument, and how in the process that follows all the group's energies are directed towards bringing everyone onside. The resulting unanimity is not, then, an illusion but a concrete testament to the group's shared understandings and commitments.

The groups in this study manifest qualities that are likely to enhance performance because, in addition to being cohesive, they are all small and demonstrably task-oriented. Yet, for all these advantages, the evidence points to a complex texture of interaction in which the tensile strength of group solidarity is tested against daily professional challenges in such a way as to bind the group closer to its own shared interpretation of the way things are and, more interestingly, the way things have to be. In such a world, 'performance' may be as much a matter of reinforcing group solidarity as delivering whatever outcomes might be professionally appropriate.

While it would be a mistake to overplay the darker side of what is an essentially positive orientation, exchanges within these groups provide some interactional evidence in support of Janis's theory. I have already mentioned the ways arguments are constructed so as to reinforce agreement rather than explore alternatives, especially when these come from outside the group, and added to this is the evidence that outsiders are constructed in ways that are more likely to suit the group's purposes than expose possible weaknesses or shortcomings. The most convincing single case was the Pen group's construction of talk so as to deflect criticism before it had even been introduced (Chapter 7), but more corrosive in the long term is the cumulative effect of mutually reinforcing positions, whether constructed indirectly through metaphor or explicitly in the form of emphatic ratification. The 'pull to uniformity' that underlies the disastrous decisions reached through groupthink is all too evident in the interaction of these groups, and the language that promotes it might, following Janis, be described as *grouptalk*.

Grouptalk is less immediately evocative than 'groupspeak', but it captures the essence of what is not an entirely negative situation, in addition to being lexically more appropriate – 'groupspeak' would imply the existence of an underlying legitimate version upon which the 'speak' is based. Grouptalk represents aspects of a group's talk that serve to reinforce the identity of the group, however this might be achieved. It might, for example, involve no more than the reinforcement of ways of speaking characteristic of the group; it could be realised through the exercise of the group's power through interaction, as in the case of Helen's attempt to act as Chair; or in its most extreme form it might serve to undermine or distort the development of ordinary activities, as it did in the Pen group's response to Nigel's criticism. In its positive manifestations it serves to reinforce group solidarity and hence encourage a shared commitment that contributes to the success of group projects, while its more negative effects might encourage a form of collective myopia that could isolate the group from outside criticism or support.

Apart from the inescapable fact that none of these otherwise very successful groups has survived, there is no direct evidence of damage caused by this aspect of group interaction even though its dangers are evident enough. What grouptalk does provide, however, is direct interactional evidence in support of the existence of groupthink. When this is combined with Janis's retrospective and largely indirect evidence of the phenomenon, the case for the existence of groupthink is strengthened. Eventually, perhaps, grouptalk could be built into a fuller picture of the model by adding to Janis's list of symptoms the following: 'the development of ways of interacting that encourage and reinforce these orientations'. As part of the larger picture, this might also explain some of the apparently contradictory findings that have arisen from further explorations of Janis's model, since grouptalk is not necessarily a negative phenomenon and it is hard to judge where the line between its positive and negative aspects is to be drawn.

Improving talk

Grouptalk and groupthink, damaging though they may be, need to be seen within a much broader and more positive context in which a secure group identity enhances the pursuit and achievement of legitimate professional business. The aim of any developmental intervention should be to build on this. Work on group interaction will therefore be only a part of a broader strategy, but nevertheless an important part, and in this section I sketch out possible approaches to this. First, though, I draw attention to a characterisation of working groups that seems to capture the spirit of the three in this study.

The importance of working groups in organisations is reflected in the sheer weight of research that has been dedicated to understanding them, and some have argued (e.g. Drucker, 1993) that organisations in the developing global economy will be team-based (for a brief overview of different views of the value of teams and team discourse, see Tietze et al., 2003: 82–3). In this context, the dimensions of my own study appear very limited, but it does fit surprisingly well with a concept that is much more ambitious in its scope and that offers a wider context within which the findings here can be positioned.

Wenger's concept of a *community of practice* resists concise definition, though the name itself captures much. Central to it is the idea of practice, represented as embedded in 'a historical and social context that gives structure and meaning to what we do' (1998: 47). Practice, argues Wenger, is a source of coherence for the relevant community, and he identifies three dimensions of the relation through which this is achieved (1998:73): mutual engagement, a joint enterprise and a shared repertoire (routines, words, ways of doing things, stories, etc.). Identity, the work of which he describes as 'ongoing', is also central to this conception and his characterisation of it might be applied equally well to the interactions analysed here:

> Identity in practice is defined socially not merely because it is reified in a social discourse of the self and social categories, but also because it is produced as a lived experience of participation in specific communities. What narratives, categories, roles and positions come to mean as an experience of participation is something that must be worked out in practice.
>
> (1998: 151)

The benefits of the concept of a community of practice from an analytical perspective have already been articulated by Holmes and Meyerhoff, who draw attention to 'regular and mutually defining interaction' (1999: 179) as a necessary condition, but I present it as a broader conceptual context within which my own study should be set. It would be possible to list coincidences between Wenger's general statements and specific evidence from my own analysis, but a single example will serve to illustrate their complementarity. On Wenger's list of indicators for the existence of a community of practice is the 'absence of introductory preambles, as if conversations and interactions were merely the continuation of an ongoing process' (1998: 125), an interesting claim but one for which he provides no direct evidence. However, the notion of a *sanctioned topic* introduced in Chapter 5 might be regarded as a paradigm case of what Wenger has in mind.

In fact, the only difference between my characterisation of the three groups and Wenger's description is a terminological issue arising from a different breadth of focus. Wenger distinguishes participation from collaboration because he takes the latter to involve only one kind of relationship (1998: 56), and in these terms my groups might be seen as communities that are in some sense deviant. However, I use the term 'collaborative' to describe these groups not because this is the only relationship that obtains within them but because it is the dominant relationship, so it captures something distinctive about their nature. This is not incompatible with Wenger's position, it merely reflects my interest in a particular *type* of community of practice.

If, as Wenger repeatedly insists, communities of practice are 'organisational assets', any attempt to improve their performance represents a sound investment. It is therefore interesting to consider that, as far as improving professional interaction is concerned, frontstage encounters have received the lion's share of attention with a residue of interest filtering through to formal backstage events, though rarely if ever to the more informal exchanges through which communities of practice develop. From a training perspective concentration on the frontstage brings its own dangers, as Cameron has shown, for where companies seek to regulate and codify the language of their employees the result may be 'regimes of verbal hygiene which seek to regulate, standardize or "improve"' communicative behaviour' (2000: 5). However, if nothing else, this at least reflects an interest in such encounters that has so far barely penetrated to the territory covered in this book.

When applied to backstage talk, Cameron's own advice to celebrate variety and complexity, or simple prescriptions such as 'Do whatever you can to increase the variety of the language with which you work' (Weick 1995: 1996), will not do. Neither will straightforward recipes for success (e.g. Robbins and Finlay, 2000: 135–44). The interactional situation is complex, and the variety within grouptalk is not all rose-tinted: the simple fact is that some varieties are best avoided. Training must therefore steer an exploratory course between the Scylla of excessive regimentation and the Charybdis of indulgent licence, perhaps working with a selection from, or combination of, the following three different approaches.

Awareness-raising

At the end of my time at the Pen I kept a promise made in the process of negotiating entry: I shared with the teachers some of the insights my research had generated. The one that had the greatest impact was my description, with examples, of the ways they closed out external criticisms or perspectives that were not consonant with their own. The group was surprised and clearly dismayed that they had fallen into such routines and discussed ways in which they might be avoided. A very simple response, for example, might be to use metalanguage in order to highlight the nature of some moves, so that a criticism would be announced as such and then discussed constructively. There is evidence that this approach can be very successful in training contexts. Pålshaugen's (2001) work on dialogue conferences, for example, aims to raise awareness of discourse as part of a broader set of strategies for promoting successful outcomes.

Exploring new styles

IEC adopted a finer grained approach, based on Edge (1992, 2002), in which they set aside an hour every week for co-operative development sessions in which strict rules for interaction were followed (see Mann, 2002 for details). A single Speaker and a number of Understanders followed a pattern in which a limited number of discourse moves by the latter were permitted in order to allow adequate interactional space for the Speaker to express and develop their ideas. Because these sessions were set aside from the normal interaction of the group, they did not interfere with its ordinary workings, perhaps even providing for some members a source of relief from the knockabout nature of everyday exchanges. There was universal support for these voluntary sessions and they were very well attended, offering colleagues a chance to explore topics that might otherwise have been neglected. Perhaps they would not appeal to all groups, and some organisations might not be willing to commit an hour a week to this enterprise, even though its facilitation of creative thinking can produce beneficial outcomes. However, it does represent an additional option that might serve some groups well.

Sensitivity training

The approach adopted by Rehling (2004) lies somewhere between the two already described. Drawing loosely on the work of Tannen and working with a broad distinction between 'overlapping' and 'turn-taking' styles, Rehling uses a number of self- and other-awareness practices to improve sensitivity to the nature of interaction. This builds on simple awareness-raising exercises by introducing participants to basic analysis and allowing the possibility of an exploration of their own talk, a process that can be extended naturally as sensitivity increases.

Set against the complexity and subtlety of professional interaction, these approaches emerge as blunt instruments, however carefully fashioned, and without a shared commitment to change on the part of group members they stand as helpless in the face of unstated resistance as Helen was in her fruitless attempt to kick-start a weekly meeting. But such commitment can emerge only from understanding, and this in turn depends on the careful excavation of everyday professional interaction.

In some ways this is a strange landscape, its shaping forces lost in a prehistory of engagements neither recorded nor remembered and the resulting interactional forms, however unusual, invisibly part of the everyday environment. Yet these features provide clues to the ways in which particular groups realise their identity through everyday engagement with the professional world, and in order to improve the quality of that engagement we must first understand its interactional particularities. This book represents a tentative early step in that direction.

References

Aldag, R.J. and Fuller, S.R. 1993. Beyond fiasco: A reappraisal of the group-think phenomenon and a new model of group decision processes. *Psychological Bulletin*, 115 (3): 533–52.

Antaki, C. 1994. *Explaining and Arguing: Social Organization of Accounts*. London: Sage.

Antaki, C. and Widdicombe, S. 1998. *Identities in Talk*. London: Sage.

Arendt, H. 1970. *On Violence*. New York: Harcourt Brace.

Arminen, I. 2000. On the context sensitivity of institutional interaction. *Discourse & Society*, 11(4): 435–58.

Atkinson, M. 1984. *Our Masters' Voices*. London: Methuen.

Atkinson, P. and Silverman, D. 1997. Kunendra's *Immortality*: The interview society and the invention of the self. *Qualitative Inquiry*, 3(3): 304–25.

Attardo, S. 2003. Introduction: the pragmatics of humor. *Journal of Pragmatics*, 35(9): 1287–94.

Bargiela-Chiappini, F. and Harris, S.J. 1997. *Managing Language: The Discourse of Corporate Meetings*. Amsterdam: John Benjamins.

Baron, R.S. and Kerr, N.L. 2003. *Group Process, Group Decision, Group Action*, Second Edition. Buckingham: Open University Press.

Beach, W.A. 1993. Transitional regularities for 'casual' "Okay" usages. *Journal of Pragmatics*, 19(4): 325–52.

Besnier, N. 1994. Involvement in linguistic practice: an ethnographic appraisal. *Journal of Pragmatics*, 22: 279–99.

Billig, M. 1987. *Arguing and Thinking: A Rhetorical Approach to Social Psychology*. Cambridge: Cambridge University Press.

Blair, J.A. 1987. Argumentation, inquiry and speech act theory. In F. van Eemeren, R. Grootendorst, J.A. Blair and C.A. Willard (eds), *Argumentation: Across the Lines of Discipline*, pp. 189–200. Dordrecht: Foris.

Boden, D. 1994. *The Business of Talk*. Cambridge: Polity Press.

Boden, D. 1995. Agendas and arrangements: everyday negotiation in meetings. In A. Firth (ed.), *The Discourse of Negotiation: Studies of Language in the Workplace*, pp. 83–99. Oxford: Pergamon.

Bolinger, D. 1975. *Aspects of Language*. 2nd Edition. New York: Harcourt, Brace, Jovanovich.

Bonner, H. 1959. *Group Dynamics: Principles and Applications*. New York: Ronald.

Boyle, R. 2000. Whatever happened to preference organisation? *Journal of Pragmatics*, 32(5): 583–604

Brown, P. and Levinson, S. 1978/1987. *Politenesss: Some Universals in Language Use*. Cambridge: Cambridge University Press.

Burns, T. 1992. *Erving Goffman*. London: Routledge.

Cameron, D. 2000. *Good to Talk? Living and Working in a Communication Culture*. London: Sage.

Cameron, L. 2003. *Metaphor in Educational Discourse*. London: Continuum.

Cameron, L. and Low, G. (eds) 1999. *Researching and Applying Metaphor*. Cambridge: Cambridge University Press.

Cicourel, A.V. 1968. *The Social Organization of Juvenile Justice*, New York: Wiley.

Cicourel, A.V. 2003. On contextualizing applied linguistic research in the workplace. *Applied Linguistics*, 24(3): 360–73.

Cooren, F. 2004. The communicative achievement of collective minding: Analysis of board meeting excerpts. *Management Communication Quarterly*, 17(4): 517–51.

Cox, J.R. and Willard C.A. (eds) 1982. *Advances in Argumentation Theory and Research*. Carbondale and Edwardsville: Southern Illinois University Press.

Craib, I. 1998. *Experiencing Identity*. London: Sage.

Craig, R.T. and Tracey, K. 2003. 'The issue' in argumentation practice and theory. In F.H. van Eemeren, J.A. Blair, C.A. Willard and A.F.S. Henkemans (eds), *Proceedings of the Fifth Conference of the International Society for the Study of Argumentation*, pp. 213–18. Amsterdam: International Society for the Study of Argumentation.

Crook, C. 1984. *Computers and the Collaborative Experience of Learning*. London: Routledge.

Crosling, G. and Ward, I. 2002. Oral communication: the workplace needs and uses of business graduate employees. *English for Specific Purposes*, 21: 41–57.

Cutting, J. 1999. The grammar of the in-group code. *Applied Linguistics*, 20(2): 179–202.

Cutting, J. 2000. *Analysing the Language of Discourse Communities*. Amsterdam: Elsevier.

De Capua, A. and Dunham, J.F. 1993. Strategies in the discourse of advice. *Journal of Pragmatics*, 20(6): 519–31.

Diamond, J. 1996. *Status and Power in Verbal Interaction: A Study of Discourse in a Close-Knit Social Network*. Amsterdam: Benjamins.

Díaz, F., Antaki, C. and Collins, A.F. 1996. Using completion to formulate a statement collectively. *Journal of Pragmatics*, 26(4): 525–42.

Dickerson, P. 2000. 'But I'm different to them': Constructing contrasts between self and others in talk-in-interaction. *British Journal of Social Psychology*, 39: 381–98.

Donato, R. 2004. Aspects of collaboration in pedagogic discourse. *Annual Review of Applied Linguistics*, 24: 284–302.

Donnellon, A. 1996. *Team Talk: The Power of Language in Team Dynamics*. Boston Mass.: Harvard Business School Press.

Drew, P. 2003. Comparative analysis of talk-in-interaction in different institutional settings: a sketch. In P.J. Glenn, C.D. LeBaron, J. Mandelbaum (eds), *Studies in Language and Social Interaction*, pp. 293–308. Mahwah, NJ: Lawrence Erlbaum.

Drew, P. and Heritage, J. 1993. Analyzing talk at work: an introduction. In P. Drew and J. Heritage (eds), *Talk at Work: Interaction in Institutional Settings*, pp. 3–65. Cambridge: Cambridge University Press.

Drew, P. and Sorjonen, M.-L. 1997. Institutional dialogue. In T.A. van Dijk (ed.), *Discourse as Social Interaction*, pp. 92–118. London: Sage.

Drucker, P.F. 1993. *Post-Capitalist Society*. New York: Harper.

Duncan, W.J. 1984. Perceived humor and social network patterns in a sample of task-oriented groups: a reexamination of prior research. *Human Relations*, 37(11): 895–907.

Duranti, A. 1986. The audience as co-author: An introduction. *Text*, 6 (3): 239–47.

Edge, J. 1992. *Cooperative Development*. Harlow: Longman.

Edge, J. 2002. *Continuing Professional Development: A Discourse Framework for Individuals as Colleagues*. Ann Arbor: University of Michigan Press.

Edwards, 1997. *Discourse and Cognition*. London: Sage.

Esser, J.K. 1998. Alive and well after 25 years: a review of groupthink research. *Organizational Behavior and Human Decision Processes*, 73(2/3): 116–41.

Fais, L. 1994. Conversation as collaboration: Some syntactic evidence. *Speech Communication*, 15: 231–42.

Fasulo, A. and Zucchermaglio, C. 2002. My selves and I: identity markers in work meeting talk. *Journal of Pragmatics*, 34(9): 1119–44.

Fillmore, C.J. 1994. Humor in academic discourse. In A. Grimshaw (ed.), *What's Going on Here? Complementary Studies of Professional Talk*, pp. 271–310. Norwood, NJ: Ablex.

Fletcher, J.K. 1999. *Disappearing Acts: Gender, Power, and Relational Practice at Work*. Cambridge, Mass: MIT Press.

Fludernik, M. 1991. The historical present tense yet again: Tense switching and oral storytelling. *Text*, 11(3): 365–97.

Forsyth, D.R. 1990. *Group Dynamics*, Second Edition. Pacific Grove, CA: Brooks/Cole.

Frey, L.R. 1988. Meeting the challenges posed during the 70s: A critical review of small group research during the eighties. Paper presented at the meeting of the Speech Communication Association, Chicago, November 1988.

Frey, L.R. 1994. The naturalistic paradigm: Studying small groups in the postmodern era. *Small Group Research*, 25: 551–77.

Frey, L.R. 1996. Remembering and 're-membering': A history of theory and research on communication and group decision making. In R.Y. Hirokawa and M.S. Poole (eds), *Communication and Group Decision Making* (2nd Edition), pp. 19–51. Thousand Oaks: Sage.

Fuller, S.R. and Aldag, R.J. 1998. Organizational Tonypandy: Lessons from a quarter of a century of the groupthink phenomenon. *Organizational Behavior and Human Decision Processes*, 73(2/3): 163–84.

Gafaranga, J. and Britten, N. 2005. Talking an institution into being: the opening sequence in general practice consultations. In K. Richards and P. Seedhouse (eds), *Applying Conversation Analysis*, pp. 75–90. Basingstoke: Palgrave Macmillan.

Garfinkel, H. 1967 *Studies in Ethnomethodology*. Englewood Cliffs, NJ: Prentice-Hall. Paperback edition 1984, *Studies in Ethnomethodology*, Cambridge: Polity Press.

Garfinkel, H. 1974. On the origins of the term 'ethnomethodology'. In R. Turner (ed.), *Ethnomethodology*, pp. 15–18. Harmondsworth: Penguin.

Georgakopoulou, A. 2001. Arguing about the future: On indirect disagreements in conversations. *Journal of Pragmatics*, 33(12): 1881–1900.

Giddens, A. 1979. *Central Problems in Social Theory: Action, Structure, and Contradiction in Social Analysis*. Basingstoke: Macmillan.

Glenn, P. 2003. *Laughter in Interaction*. Cambridge: Cambridge University Press.

Goffman, E. 1959/71. *The Presentation of Self in Everyday Life*. Harmondsworth: Penguin Books. Originally published in 1959 by Anchor Books.

Goffman, E. 1967. *Interaction Ritual: Essays on Face to Face Behavior*. New York: Anchor Books.

Goffman, E. 1981. *Forms of Talk*. Oxford: Blackwell.

Gomm, R., Hammersley, M. and Foster, P. 2000. Case study and generalization. In R. Gomm, M. Hammersley and P. Foster (eds), *Case Study Method*, pp. 98–115. London: Sage.

Goodwin, C. 1986. Audience diversity, participation and interpretation. *Text*, 6(3): 283–316.

Goodwin, C. and Duranti, A. 1992. Rethinking context: an introduction. In A. Duranti and C. Goodwin (eds), *Rethinking Context: Language as an Interactive Phenomenon*, pp. 1–42. Cambridge: Cambridge University Press.

Goodwin, G. and Goodwin, M.H. 1990. Interstitial argument. In A.D. Grimshaw (ed.), *Conflict Talk*, pp. 85–117. Cambridge: Cambridge University Press.

Goodwin, M.H. and Goodwin, C. 1987. Children's arguing. In S. Philips, S. Steele and C. Tanz (eds), *Language, Gender, and Sex in Comparative Perspective*, pp. 240–48. Cambridge: Cambridge University Press.

Gordon, C. 2003. Aligning as a team: Forms of conjoined participation in (step-family) interaction. *Research on Language and Social Interaction*, 36(4): 395–431.

Grice, H.P. 1975. Logic and conversation. In P. Cole and J.L. Morgan (eds), *Syntax and Semantics 3: Speech Acts*, pp. 41–58. New York: Academic Press.

Gruber, H. 1998. Diagreeing: Sequential placement and internal structure of disagreements in conflict episodes. *Text*, 18(4): 467–503.

Gumperz, J.J. 1982. *Discourse Strategies*. Cambridge: Cambridge University Press.

Hammersley, M. 1980. A peculiar world? Teaching and learning in an inner-city school. Unpublished PhD thesis, University of Manchester.

Hammersley, M. 1981. Ideology in the staffroom? A critique of false consciousness. In L. Barton and S. Walker (eds), *Schools, Teachers and Teaching*, pp. 331–42. Lewes: Falmer Press.

Hammersley, M. 1984. Staffroom news. In A. Hargreaves and P. Woods (eds), *Classrooms and Staffrooms: The Sociology of Teachers and Teaching*, pp. 203–14. Milton Keynes: Open University Press.

Handelman, D. and Kapferer, B. 1972. Forms of joking: a comparative approach. *American Anthropologist*, 74: 484–517.

Hannay, L.M. 1996. The role of images in the secondary school change process. *Teachers and Teaching: Theory and Practice*, 2(1): 105–21.

Hargreaves, A. 1992. Cultures of teaching: a focus for change. In A. Hargreaves and A.G. Fullan (eds), *Understanding Teacher Development*, pp. 216–40. London: Cassell.

Hargreaves, A. 1994. *Changing Teachers, Changing Times*. London: Cassell.

Hargreaves, D.H. 1977. The process of typification in the classroom. *British Journal of Educational Psychology*, 47: 274–84.

Heritage, J. 1984a. A change-of-state token and aspects of its sequential placement. In J.M. Atkinson and J. Heritage (eds), *Structures of Social Action*, pp. 299–345. Cambridge: Cambridge University Press.

Heritage, J. 1984b. *Garfinkel and Ethnomethodology*. Cambridge: Polity Press.

Heritage, J.C. and Watson, D.R. 1979. Formulations as conversational objects. In G. Psathas (ed.), *Everyday Language*, pp. 123–62. New York: Wiley.

Hertzler, J.O. 1970. *Laughter: A Socio-scientific Analysis*. New York: Exposition Press.

Hester, S. 1996. Laughter in its place. In G. Paton, C. Powell and S. Wagg (eds), *The Social Faces of Humour: Issues and Practices*, pp. 245–69. Aldershot: Arena Press.

Hester, S. and Eglin, P. 1997. Membership categorization analysis: An introduction. In S. Hester and P. Eglin (eds), *Culture in Action: Studies in Membership Categorization Analysis*, pp. 1–23. Washington, DC: International Institute for Ethnomethodology and Conversation Analysis, and University Press of America.

Holmes, J. 1998. No joking matter! The functions of humour in the workplace. *Australian Linguistic Society ALS98 Papers*. http://www.cltr.uq.edu.au/als98/holme358.html

Holmes, J. 2000a. Doing collegiality and keeping control at work: Small talk in government departments. In J. Coupland (ed.), *Small Talk*, pp. 32–61. Harlow: Pearson Education.

Holmes, J. 2000b. Politeness, power and provocation: how humour functions in the workplace. *Discourse Studies*, 2(2): 159–85.

Holmes, J. and Marra, M. 2002. Having a laugh at work: How humour contributes to workplace culture. *Journal of Pragmatics*, 34(12): 1683–1710.

Holmes, J. and Marra, M. 2004. Relational practice in the workplace: Women's talk or gendered discourse? *Language in Society*, 33: 377–98.

Holmes, J. and Meyerhoff, M. 1999. The community of practice: Theories and methodologies in language and gender research. *Language in Society*, 28: 173–83.

Holmes, J. and Stubbe, M. 2003. *Power and Politeness in the Workplace*. London: Longman.

Hutchins, E. 1995. *Cognition in the Wild*. Cambridge, MA: MIT Press.

Hymes, D. 1986. Models of the interaction of language and social life. In J.J. Gumperz and D. Hymes (eds), *Directions in Sociolinguistics*, pp. 35–71. Oxford: Blackwell.

Iedema, R. and Wodak, R. 1999. Introduction: organizational discourses and practices. *Discourse & Society*, 10(1): 5–19.

Iedema, R. and Scheeres, H. 2003. From doing work to talking work: Renegotiating knowing, doing, and identity. *Applied Linguistics*, 24(3): 316–37.

Jacobs, S. and Jackson, S. 1981. Conflict in collaborative decision-making. In N.R. Blyler and C. Thralls (eds), *Professional Communication: The Social Perspective*, pp. 144–62. Newbury Park, CA: Sage.

Jacobs, S. and Jackson, S. 1982. Conversational argument: A discourse analytic approach. In J.R. Cox and C.A. Willard (eds), pp. 205–57.

Janis, I.L. 1972. *Victims of Groupthink*. Boston: Houghton Mifflin.

Janis, I.L. 1982. *Groupthink* (2nd ed.). Boston: Houghton Mifflin.

Janis, I.L. 1989. *Crucial Decisions: Leadership in Policymaking and Crisis Management*. New York: Free Press.

Jefferson, G. 1972. Side sequences. In D. Sudnow (ed.), *Studies in Social Interaction*, pp. 294–338. New York: Free Press.

Jefferson, G. 1978. Sequential aspects of storytelling in conversation. In J. Schenkein (ed.), *Studies in the Organization of Conversational Interaction*, pp. 219–48. New York: Academic Press.

Jefferson, G. 1979. A technique for inviting laughter and its subsequent acceptance/declination. In: G. Psathas (ed.), *Everyday Language: Studies in Ethnomethodology*, pp. 79–96. New York: Irvington.

Jefferson, G. 1983. Notes on the systematic deployment of the acknowledgement tokens 'Yeah' and 'mm hm'. *Tilberg Papers in Language and Literature*, 28: 1–18. Tilberg, The Netherlands: Tilberg University. http://www.liso.ucsb.edu/Jefferson/Acknowledgment

•

Jefferson, G. 1984. On the organization of laughter in talk about troubles. In J.M. Atkinson and J. Heritage (eds), *Structures of Social Action*, pp. 346–69. Cambridge: Cambridge University Press.

Jefferson, G. 1985. An exercise in the transcription and analysis of laughter. In T. Van Dijk (ed.), *Handbook of Discourse Analysis Vol. 3: Discourse and Dialogue*, pp. 25–34. London: Academic Press.

Jefferson, G. 2004. A note on laughter in 'male–female' interaction. *Discourse Studies*, 6(1): 117–33.

Jefferson, G., Sacks, H. and Schegloff, E.A. 1987. Notes on laughter in the pursuit of intimacy. In: G. Button and J.R.E. Lee (eds), *Talk and Social Organisation*, pp. 152–205. Clevedon: Multilingual Matters.

Kainan, A. 1992. Themes of individualism, competition, and cooperation in teachers' stories. *Teaching and Teacher Education*, 8(5/6): 441–50.

Kangasharju, H. 1996. Aligning as a team in multiparty conversation. *Journal of Pragmatics*, 26(3): 291–319.

Kangasharju, H. 2002. Alignment in disagreement: forming oppositional alliances in committee meetings. *Journal of Pragmatics*, 34(10-11): 1447–71.

Keller, E.F. 1995. *Refiguring Life: Metaphors of Twentieth-Century Biology*. New York: Columbia University Press.

Kleiner, B. 1998. Whatever – Its use in 'pseudo-argument'. *Journal of Pragmatics*, 30(5): 589–613.

Koester, A.J. 2004. Relational sequences in workplace genres. *Journal of Pragmatics*, 36(8): 1405–28.

Kotthoff, H. 1993. Disagreement and concession in disputes: On the context sensitivity of preference structures. *Language in Society*, 22: 193–216.

Labov, W. 1972/1977. *Language in the Inner City*. Oxford: Blackwell. (Originally published in 1972 by University of Pennsylvania Press.)

Labov, W. and Waletzky, J. 1967. Narrative analysis: oral versions of personal experience. In J. Helms (ed.), *Essays on the Verbal and Visual Arts*, pp. 12–44. Seattle: University of Washington Press.

Lakoff, G. and Johnson, M. 1980. *Metaphors We Live By*. Chicago Ill.: University of Chicago Press.

Larrue, J. and Trognon, A. 1993. Organization of turn-taking and mechanisms for turn-taking repairs in a chaired meeting. *Journal of Pragmatics*, 19(2): 177–96.

Leana, C.R. 1985. A partial test of Janis' Groupthink model: Effects of group cohesiveness and leader behaviour on defective decision making. *Journal of Management*, 11(1): 5–17.

Lepper, G. 2000. *Categories in Text and Talk*. London: Sage.

Lerner, G.H. 1991. On the syntax of sentences-in-progress. *Language in Society*, 20: 441–58.

Lerner, G.H. 1992. Assisted storytelling: Deploying shared knowledge as a practical matter. *Qualitative Sociology*, 15: 247–71.

Lerner, G.H. 1996. On the 'semi-permeable' character of grammatical units in conversation: conditional entry into the turn space of another speaker. In E. Ochs, E.A. Schegloff and S.A. Thompson (eds), *Interaction and Grammar*, pp. 238–76. Cambridge: Cambridge University Press.

Lerner, G.H. 2002. Turn-sharing: the choral co-production of talk-in-interaction. In C.E. Ford, B.A. Fox and S.A. Thompson (eds), *The Language of Turn and Sequence*, pp. 225–56. Oxford: Oxford University Press.

Leuder, I. and Antaki, C. 1988. Completion and dynamics in explanation seeking. In C. Antaki (ed.), *Analysing Everyday Explanation: A Casebook of Methods*, pp. 145–55. London: Sage.

Levinson, S.C. 1983. *Pragmatics*. Cambridge: Cambridge University Press.

Linstead, S. 1988. 'Jokers wild': Humour in organisational culture. In C. Powell and E.C. Paton (eds), *Humour in Society: Resistance and Control*, pp. 123–48. Basingstoke: Macmillan.

Little, J.W. 1990. The persistence of privacy: Autonomy and initiative in teachers' professional relations. *Teachers' College Record*, 9(4): 509–36.

Long, D.L. and Graessner, A.C. 1988. Wit and humour in discourse processing. *Discourse Processes*, 11: 35–60.

Longley, J. and Pruitt, D.G. 1980. Groupthink: A critique of Janis' theory. In L. Wheeler (ed.), *Review of Personality and Social Psychology*, pp. 507–13. Newbury Park, CA: Sage.

MacLure, M. 1993. Mundane autobiography: Some thoughts on self-talk in research contexts. *British Journal of Sociology of Education*, 14(4): 373–84.

Makan, J.M. and Marty, D.L. 2001. *Cooperative Argumentation*. Long Grove, Ill.: Waveland Press.

Malone, M.J. 1997. *Worlds of Talk*. Cambridge: Polity Press.

Mandelbaum, J. 1989. Interpersonal activities in conversational storytelling. *Western Journal of Speech Communication*, 53: 114–26.

Mandelbaum, J. 1993. Assigning responsibility in conversational storytelling: The interactional construction of reality. *Text*, 13(2): 247–66.

Mann, S. 2002. The development of discourse in a discourse of development: A case study of a group constructing a new discourse. Unpublished PhD thesis, University of Aston in Birmingham.

Maturana, H.R. and Varela, F.J. 1998. *The Tree of Knowledge: The Biological Roots of Human Understanding*. Boston and London: Shambhala.

Maynard, C. 1985. How children start arguments. *Language in Society*, 14: 1–30.

Maynard, D.W. 1987. Introduction. *Social Psychology Quarterly* Special issue: Language and Social Interaction, 50(2): v–vi.

McCauley, C. 1989. The nature of social influence in groupthink: Compliance and internalization. *Journal of Personality and Social Psychology*, 57: 250–60.

McGrath, J.E. and Altermatt, T.W. 2001. Observation and analysis of group interaction over time: some methodological and strategic choices. In M.A. Hogg and R.S. Tindale (eds) *Blackwell Handbook of Social Psychology: Group Processes*, pp. 525–56. Oxford: Blackwell.

McMartin, C., Wood, L.A. and Kroger, R.O. 2001. Facework. In P. Robinson and H. Giles (eds), *The New Handbook of Language and Social Psychology*, pp. 221–37. Chichester: John Wiley & Sons.

Mercer, N. 2000. *Words and Minds: How We Use Language to Think Together*. London: Routledge.

Miller, G. 1994. Towards ethnographies of institutional discourse. *Journal of Contemporary Ethnography*, 23(3): 280–306.

Miller, G. 1997. Towards ethnographies of institutional discourse: proposals and suggestions. In G. Miller and R. Dingwall (eds), *Context and Method in Qualitative Research*, pp. 155–71. London: Sage.

Moorhead, G. 1982. Groupthink: Hypothesis in need of testing. *Group & Organization Studies*, 7: 429–44.

Moscovici, S. 1984. The phenomena of social representations. In R.M. Farr and S. Moscovici (eds), *Social Representations*, pp. 3–69. Cambridge: Cambridge University Press.

Mulholland, J. 1996. A series of story turns: Intertextuality and collegiality. *Text*, 16(4): 535–55.

Mullen, B. and Copper, C. 1994. The relation between group cohesiveness and performance: An integration. *Psychological Bulletin*, 115(2): 210–27.

Muntigl, P. and Turnbull, W. 1998. Conversational structure and facework in arguing. *Journal of Pragmatics*, 29(3): 215–56.

Nias, J., Southworth, G. and Yeomans, R. 1989. *Staff Relationships in the Primary School: A Study of Organizational Cultures*. London: Cassell.

Norrick, N.R. 1994. Involvement and joking in conversation. *Journal of Pragmatics*, 22(3–4): 409–30.

Norrick, N.R. 2003. Issues in conversational joking. *Journal of Pragmatics*, 35(9): 1333–59.

Nwoye, O.G. 1992. Linguistic politeness and socio-cultural variations of the notion of face. *Journal of Pragmatics*, 18(4): 309–28.

O'Keefe, B.J. and Benoit, P.J. 1982. Children's arguments. In Cox and Willard (eds), pp. 154–83.

O'Keefe, D.J. 1977. Two concepts of argument. *Journal of the American Forensic Association*, 13: 121–8.

O'Keefe, D.J. 1982. The concepts of argument and arguing. In Cox and Willard (eds), pp. 3–23.

Ochs, E., Taylor, C., Rudolph, D. and Smith, R. 1992. Storytelling as a theory-building activity. *Discourse Processes*, 15: 37–72.

Pålshaugen, Ø. 2001. The use of words: Improving enterprises by improving their conversations. In P. Reason and H. Bradbury (eds), *Handbook of Action Research: Participative Inquiry and Practice*, pp. 209–18. London: Sage.

Parker, M. 2000. *Organizational Culture and Identity: Unity and Division at Work*. London: Sage.

Paton, G.E.C. 1992. *Humour at Work and the Work of Humour*. Birmingham: Aston Business School, Aston University.

Peräkylä, A. and Vehviläinen, S. 2003. Conversation analysis and the professional stocks of interactional knowledge. *Discourse & Society*, 14(6): 727–50.

Polanyi, L. 1985. *Telling the American Story: A Structural and Cultural Analysis of Conversational Storytelling*. Norwood, NJ: Ablex.

Pomerantz, A. 1984a. Agreeing and disagreeing with assessment: Some features of preferred/dispreferred turn shapes. In J.M. Atkinson and J. Heritage (eds), *Structures of Social Action: Studies in Conversation Analysis*, pp. 57–101. Cambridge: Cambridge University Press.

Pomerantz, A. 1984b. Pursuing a response. In J. Atkinson and J. Heritage (eds), *Structures of Social Action: Studies in Conversation Analysis*, pp. 152–63. Cambridge: Cambridge University Press.

Poole, M.S. and Hirokawa, R.Y. 1996. Introduction: communication and group decision making. In R.Y. Hirokawa and M.S. Poole (eds), *Communication and Group Decision Making* (2nd Edition), pp. 3–18. Thousand Oaks, CA: Sage.

Poole, M.S., Seibold, D.R. and McPhee, R.D. 1996. The structuration of group decisions. In R.Y. Hirokawa and M.S. Poole (eds), *Communication and Group Decision Making* (2nd Edition), pp. 114–46. Thousand Oaks, CA: Sage.

Postmes, T. 2003. A social identity approach to communication in organizations. In S.A. Haslam, D. van Knippenberg, M.J. Platow and N. Ellemers (eds), *Social Identity at Work*, pp. 81–97. New York and Hove: Psychology Press.

Potter, J. 2000. Post-cognitive psychology. *Theory and Psychology*, 10(1): 31–7.

Potter, J. and Edwards, D. 2001. Discursive Social Psychology. In W.P. Robinson and H. Giles (eds), *The New Handbook of Language and Social Psychology*, pp. 103–18. Chichester: John Wiley & Sons.

Powell, C. 1988. A phenomenological analysis of humour in society. In C. Powell and E.C. Paton (eds), *Humour in Society: Resistance and Control*, pp. 86–105. Basingstoke: Macmillan.

Psathas, G. 1999. Studying the organization in action: Membership categorization and interaction analysis. *Human Studies*, 22: 139–62.

Putnam, L.L. 1986. Conflict in group decision-making. In R.Y. Hirokawa and M.S. Poole (eds), *Communication and Group Decision-making*, pp. 175–96. Beverly Hills: Sage.

Putnam, L.L. and Fairhurst, G.T. 2001. Discourse analysis in organizations: Issues and concerns. In F.M. Jablin and L.L. Putnam (eds), *The New Handbook of Organizational Communication*, pp. 78–136. Thousand Oaks, CA: Sage.

Rehling, L. 2004. Improving teamwork through awareness of conversational style. *Business Communication Quarterly*, 67(4): 475–82.

Richards, K. 1999. Working towards common understandings: Collaborative interaction in staffroom stories. *Text*, 19(1): 143–74.

Rist, R. 1973. *The Urban School*. Cambridge Mass.: MIT Press.

Robbins, H. and Finley, M. 2000. *Why Teams Don't Work: What Went Wrong and How to Make it Right*. London: Texere.

Rosenholtz, S.J. 1989. *Teachers' Workplace: The Social Organization of Schools*. London: Longman.

Sacks, H. 1972. An initial investigation of the usability of conversational data for doing sociology. In D. Sudnow (ed.), *Studies in Social Interaction*, pp. 31–74. New York: Free Press.

Sacks, H. 1974. An analysis of the course of a joke's telling in conversation. In R. Bauman and J. Sherzer (eds), *Explorations in the Ethnography of Speaking*, pp. 324–67. Cambridge: Cambridge University Press.

Sacks, H. 1987. On the preferences for agreement and contiguity in sequences in conversation. In G. Button and J.R.E. Lee (eds), *Talk and Social Organisation*, pp. 54–69. Clevedon: Multilingual Matters.

Sacks, H. 1992a. *Lectures on Conversation*, Vol. 1, ed. G. Jefferson. Oxford: Blackwell.

Sacks, H. 1992b. *Lectures on Conversation*, Vol. 2, ed. G. Jefferson. Oxford: Blackwell.

Sacks, H., Schegloff, E.A. and Jefferson, G. 1974. A simplest systematics for the organization of turn-taking for conversation. *Language*, 50(4): 696–735.

Sarangi, S. and Roberts, C. 1999. The dynamics of interactional and institutional orders in work-related settings. In S. Sarangi and C. Roberts (eds), *Talk, Work and Institutional Order*, pp. 1–57. Berlin: Mouton de Gruyter.

Schegloff, E.A. 1987. Between micro and macro: contexts and other connections. In J.C. Alexander, B. Giesen, R. Münch and N.J. Smelser (eds), *The Micro-Macro Link*, pp. 207–34. Berkeley: University of California Press.

Schegloff, E.A. 1991a. Conversation analysis and socially shared cognition. In L.B. Resnick, J.L. Levine and S.D. Teasley (eds), *Perspectives on Socially Shared Cognition*, pp. 150–71. Washington DC: American Psychological Association.

Schegloff, E.A. 1991b. Reflections on talk and social structure. In D. Boden and D.H. Zimmerman (eds), *Studies in Ethnomethodology and Conversation Analysis*, pp. 44–70. London: Polity Press.

Schegloff, E.A. and Sacks, H. 1973. Opening up closings. *Semiotica*, 8: 289–327.

Schenkein, J.N. 1972. Towards the analysis of natural conversation and the sense of 'heheh'. *Semiotica*, 6: 344–77.

Schiffrin, D. 1981. Tense variation in narrative. *Language*, 57(1): 45–62.

Schiffrin, D. 1984. Jewish argument as sociability. *Language in Society*, 13: 311–35.

Schiffrin, D. 1990. The management of a co-operative self during argument: The role of opinions and stories. In A.D. Grimshaw (ed.), *Conflict Talk*, pp. 241–59.

Sillince, J.A.A. 1995. Shifts in focus and scope during argumentation. *Journal of Pragmatics*, 24(4): 413–31.

Sinclair, J.M. and Coulthard, M. 1975. *Towards an Analysis of Discourse*. Oxford: Oxford University Press.

Smircich, L. and Stubbart, C. 1985. Strategic management in an enacted world. *Academy of Management Review*, 10: 724–36.

Smithson, J. and Diaz, F. 1996. Arguing for a collective voice: Collaborative strategies in problem-oriented conversation. *Text*, 16 (2), 251–68.

Sollitt-Morris, L. 1997. Taking a break: Humour as a means of enacting power in asymmetrical discourse. *Language, Gender and Sexism*, 7(2): 81–103.

Spencer-Oatey, H. 2002. Managing rapport talk: Using rapport sensitive incidents to explore the motivational concerns underlying the management of relations. *Journal of Pragmatics*, 34(5): 529–45.

Stephenson, R.M. 1951. Conflict and control functions of humour. *American Journal of Sociology*, 56: 569–74.

Stohl, C. and Putnam, L.L. 2003. Communication in bona fide groups: A retrospective and prospective account. In L.R. Frey (ed.), *Group Communication in Context: Studies of Bona Fide Groups*, pp. 399–414. Mahwah, NJ: Lawrence Erlbaum.

Stubbe, M. 1998. Researching language in the workplace: a participatory model. Australian Linguistic Society Papers 1998. Accessible online at: http://www.cltr.uq.edu.au/als98/stubb266.html

Stubbe, M., Lane, C., Hilder, J., Vine, E., Vine, B., Marra, M., Holmes. J. and Weatherall, A. 2003. Multiple discourse analyses of a workplace interaction. *Discourse Studies*, 5(3): 351–88.

Stubbs, M. 1992. Institutional linguistics: Language and institutions, linguistics and sociology. In M. Pütz (ed.),*Thirty Years of Linguistic Evolution*, pp. 189–211. Duisberg: University of Duisberg.

Tannen, D. 1984. *Conversational Style: Analyzing Talk among Friends*. Norwood, NJ: Ablex.

Tannen, D. 1989. *Talking Voices: Repetition, Dialogue, and Imagery in Conversational Discourse*. Cambridge: Cambridge University Press.

Tetlock, P.E. 1979. Identifying victims of groupthink from public statements of decision makers. *Journal of Personality and Social Psychology*, 37: 1314–24.

Tetlock, P.E., Peterson, R.S., McGuire, C., Chang, S. and Feld, P. 1992. Assessing political group dynamics: A test of the groupthink model. *Journal of Personality and Social Psychology*, 63: 403–25.

Thomas, S.G. 1995. Preparing business students more effectively for real-world communication. *Journal of Business and Technical Communication*, 9(4): 461–74.

Thornbury, S. 1991. Metaphors we work by: EFL and its metaphors. *ELT Journal*, 45(3): 193–200.

Tietze, S., Cohen, L. and Musson, G. 2003. *Understanding Organizations through Language*. London: Sage.

Tindale, R.S., Meisenhelder, H.M., Dykema-Engblade, A.A. and Hogg, M.A. 2001. Shared cognition in small groups. In M.A. Hogg and R.S. Tindale (eds), *Blackwell Handbook of Social Psychology: Group Processes*, pp. 1–30. Oxford: Blackwell.

Toolan, M.J. 1989. *Narrative: A Critical Linguistic Introduction*. London: Routledge.

Tracey, K. and Aschcroft, C. 2001. Crafting policies about controversial values: How wording disputes manage a group dilemma. *Journal of Applied Communication Research*, 29(4): 297–316.

Tracey, K. and Carjuzá, J. 1993. Identity enactment in intellectual discussion. *Journal of Language and Social Psychology*, 18(3): 171–94.

Tracy, K. and Naughton, J.M. 2000. Institutional identity-work: a better lens. In J. Coupland (ed.), *Small Talk*, 62–83. Harlow: Longman.

Turniansky, B. and Hare, A.P. 1998. *Individuals in Groups and Organizations*. London: Sage.

van Eemeren, F.H., Grootendorst, R., Jackson, S. and Jacobs, S. 1993. *Reconstructing Argumentative Discourse*. Tuscaloosa, AL.: University of Alabama Press.

van Knippenberg, D. and Ellemers, N. 2003. Social identity and group performance: identification as the key to group-oriented effort. In S.A. Haslam, D. van Knippenberg, M.J., Platow, and N. Ellemers (eds), *Social Identity at Work*, pp. 29–42. New York and Hove: Psychology Press.

Van Maanen, J. 2001. Afterword: Natives 'R' us: Some notes on the ethnography of organizations. In D.N. Gellner and E. Hirsch (eds), *Inside Organizations: Anthropologists at Work*, pp. 233–61. Oxford: Berg.

Van Vree, W. 1999. *Meetings, Manners and Civilization: The Development of Modern Meeting Behaviour*. London: Leicester University Press.

Weick, K.E. 1995. *Sensemaking in Organizations*. Thousand Oaks CA: Sage.

Weick, K.E. and Roberts, K.H. 1993. Collective mind in organizations: Heedful interrelating on flight decks. *Administrative Science Quarterly*, 38: 357–81.

Wenger, E. 1998. *Communities of Practice: Learning, Meaning and Identity*. Cambridge: Cambridge University Press.

Wennerstrom, A. 2000. Is it me or is it hot in here? Menopause, identity, and humor. *Humor*, 13(3): 313–31.

Willard, C.A. 1989. *A Theory of Argumentation*. Tuscaloosa, AL: University of Alabama Press.

Williams, M. 2002. Generalization in interpretive research. In T. May (ed.), *Qualitative Research in Action*, pp. 125–143. London: Sage.

Wilson, J. 1989. *On the Boundaries of Conversation*. Oxford: Pergamon.

Wolfson, N. 1979. The conversational historical present alternation. *Language*, 55(1): 168–82.

Wolfson, N. 1982. *The Conversational Historical Present in American English Narrative*. Cinnarminson, NJ: Foris Publications.

Woods, P. 1979. *The Divided School*. London: Routledge & Kegan Paul.

Zajdman, A. 1995. Humorous face-threatening acts: Humor as strategy. *Journal of Pragmatics*, 23(3), 325–39.

Index

Printed in the United States
221802BV00001B/39/P

9 780230 580114